The Real Secret

The Real Secret

HOW TO SOLVE YOUR PROBLEMS
—AND IMPROVE YOUR LIFE—
with Science and Reason

DAVID G. MCAFEE

Pitchstone Publishing
Durham, North Carolina

Pitchstone Publishing
Durham, North Carolina
www.pitchstonebooks.com

Library of Congress Cataloging in Publication Control Number: 2026023975

ISBN (paperback): 9781634312967
ISBN (ebook): 9781634312974

This book is dedicated to my wife, Rae, who is my favorite editor and has always encouraged me to better myself in every way possible, as well as to Stephanie Guttormson, who helped inspire the title of this book several years prior to me writing it, and to Jeffrey Quillinan, a key supporter who made this book possible. Thank you both for your contributions.

Contents

"A Clean Sweep"

"Please read pages 30 through 50 tonight, kids. I know some of you haven't been doing the reading," said Mrs. Miller, a larger-than-life figure who had already been teaching high school for 20 years. She was two months into the 21st year and knew the signs when the students weren't keeping up with their assignments.

"And don't forget Retrograde break is coming up. Three weeks of no classes, but I still expect you to do the reading," Mrs. Miller added, her greying hair falling over her black-rimmed glasses just a little. As she pushed the curls out of her eyes, she noticed a student with one hand in the air and her other hand supporting it, the universal symbol that she was frantically trying to be noticed. "Yes, Kailey?"

"Should you assign us homework over Retrograde break?" Kailey asked. "I get really bad brain fog when Mercury is in retrograde, and most of us have Vision Boards due for other classes. I don't know that I can do the required reading."

A visibly frustrated Mrs. Miller protested.

"Retrograde break is part of our society, but I've never seen concrete proof it gives 'brain fog,'" the teacher replied. After thinking for a second, she then conceded, "But, I don't think I will win this fight with the board. Very well. No reading will be assigned over the break. But I expect you to come back to class ready to learn!"

Just as Mrs. Miller uttered those words, a subdued buzzing sound signaled that class was over. She rushed to get out the last bit of what she was trying to say.

"And don't forget the reading tonight! It's 20 pages, and you won't pass tomorrow's test without that knowledge!" she half-shouted as kids were starting to leave the classroom.

As they walked by Mrs. Miller's desk on the way out of the class, the students kept their heads down, except for Kailey, who approached.

"Thank you, Mrs. Miller. As a Pisces, I'm easily distracted, so having this homework over Retrograde wouldn't be good," the girl said.

Mrs. Miller simply nodded, but the educator failed to hide her facial expression, which betrayed her disbelief. She also noticed that the student behind Kailey, Lotta, groaned at Kailey, her light freckles moving position as she contorted her face that was partially hiding behind her long brown hair, but she didn't say a thing. Still, her eyes met Mrs. Miller's as she walked toward the door.

Just as Lotta was walking out of the classroom, a second buzzer sounded. This one wasn't something that the students had remembered hearing before. Even Mrs. Miller appeared shocked. But she didn't have much time for surprises. As the buzzer was still going off, a rushed Principal Byrne appeared at the door. Barely sticking his mostly bald head through the entryway, the principal looked at Mrs. Miller and shouted in a stern tone, "Linda, get them to the multipurpose room right now."

"OK, what's going on?" Mrs. Miller asked him.

He responded with only two short words: "CLEAN SWEEP."

Mrs. Miller's eyes widened as she started shouting to the kids to meet her in the gymnasium.

"OK, kids, this is a code purple. That means gather your things and single-file your way to the multipurpose room," the teacher insisted.

Mrs. Miller herself gathered her things and followed the last of the kids through the door. They were all walking as a group through the hall, with Principal Byrne leading up front and setting the quick pace. The school's speaker system sputtered and crackled before a familiar voice boomed.

The voice was that of the Vice Principal, Jim Warren, who was trying to corral people into the school's multipurpose room, where all people from the town gathered in emergency situations.

"This is a code purple, so good vibes only, everyone," Warren's voice was heard saying. "Please stay calm, and gather in the multipurpose room, where we will prepare for Retrograde by eliminating the spirit barrier. We'll have a clean sweep this time around."

As Mrs. Miller approached the gathering spot with her students close behind her, she realized she would have to split from most of them. At the entrance of the multipurpose room was a city employee who was lighting sage and wafting the smoke into each corner of the large gym. A second town worker was strictly separating individuals into groups based on their birth information. Mrs. Miller was guided into the section for people with the Pisces sign, such as Kailey. Others followed signs that said "Aries," "Cancer," "Taurus," etc., so that everyone was placed with their astrological chart in mind. The city was so serious about this that the individuals with complementary signs were positioned near one another, such that those with the Aries sign were next to the Libras, but nowhere near the group of Capricorns.

Principal Byrne took the stage after only a few minutes and started giving what appeared to be a rehearsed speech.

"Thank you all for being here! As I'm sure you know, our society has been experiencing a rough patch. We tried harnessing all of our positive energy, but despite our best efforts, fuel and food prices are soaring, disease runs rampant, and the planet is heating up at a rate that can't possibly be sustained," he said. "In times like this, we have been called to action. There has to be a *reason* that none of our normal rituals and systems have worked. Because we don't know what that reason is, we have to cleanse."

Another teacher, Mr. Handler, ducked on stage and gave the Principal a note before silently backing away. Principal Byrne winced a little, but mostly kept it together as he continued to speak.

"Our system is not perfect, but we know that doubters and those who doubt our spiritual practices make things worse," he said softly. "We know that our rituals don't work if they aren't backed by sincere belief from across the city."

Principal Byrne paused a moment and looked out at the sea of faces in the multipurpose room.

"So, how do we find a doubter?" the Principal asked the crowd rhetorically. "This decision is never an easy one."

Many of the people in that room had no idea what would come next, but others remembered the last Clean Sweep 20 years earlier. That time, it was a skeptical teacher named Brad

Mason who was chosen. The crowd was filled with murmurs as everyone was wondering what would happen, but Mrs. Miller was staying dead silent. She knew that she had a reputation for doubting some of the popular spiritual practices of the society.

"This is a moment nobody wants to come, but it's also necessary if we want to succeed in our society," Principal Byrne said as he shuffled his feet on the stage and made eye contact with the mayor in the crowded room. "It is time."

The moment those words left the Principal's lips, all the lights in the room dimmed, except a spotlight that illuminated one grouping: Pisces. Mrs. Miller felt the heat from the light, as well as from the stares of those in the room, immediately. Her face turned the color of a Red Delicious apple, and she shrank behind her own colleagues. She had been a teacher long enough to experience two other Clean Sweeps, and knew all too well what was next for the doubter they single out, but the kids in the Pisces group appeared blissfully unaware.

The spotlight narrowed until Mrs. Miller was brightly illuminated, and she was terrified. She knew she didn't accept *all* of the rituals associated with the society, but she did her best to work within the framework.

After debating making a run for it, Mrs. Miller cleared her throat and said, "Listen, you guys. You know me. You know this isn't me. I'm not the problem here. I didn't even assign reading over Retrograde break, for the planets' sake!"

"Maybe we don't even have to do this at all!" she added. "It's possible that we can achieve our goals another way. With science, we …"

Just then, a loud blowhorn sounded and stopped Mrs. Miller's sentence in the middle of her thought.

"Quiet, Mrs. Miller," Principal Byrne said. "Step aside."

A confused Mrs. Miller did as she was told and took a step to the left, but the spotlight did not shift. It stayed right where it was, illuminating the person behind the panicked educator.

The Principal said, "Lotta Jackson, we have had reports about you not following the spirits, ignoring your crystal work, and failing to adhere to your astrological charting. You have been chosen for the Clean Sweep."

Instead of relief, Mrs. Miller felt only fear and hatred.

Lotta didn't know how to react, so she nervously grabbed onto Mrs. Miller's pant suit. The teacher shouted, "NO. You can't do this. Not her!"

Before Mrs. Miller could continue wailing, several local authorities appeared behind Lotta and began dragging her away.

"There is another way!" the teacher shouted as she saw her student being taken from her. "Our society can be better than this!"

"Code blue," the Principal said into his black walkie-talkie. "Restrain the teacher."

Two men came to grab Mrs. Miller, and they put a rag with a eucalyptus oil solution over the educator's mouth. As her vision faded, she heard the words of the Principal.

"This isn't a pretty process," the administrator said. "But it is what we have."

He continued:

"Our way of life must be preserved, and talk of science doesn't help that. We need to recognize the power of our methods, because we can't have anything dimming our collective aura. As we did many years ago, we now start this next planetary cycle with a clean sweep, free of the doubt that clouds our own spiritual power. This is the only path to get beyond the sadness, hunger, and depression that haunt our society. Right now, many in our world are crushed, often losing everything and failing to accomplish their goals, despite our focus on vision boarding and breath work. The only way forward is to purge the doubter and open up our chakras as we move toward our truth. It's science that threatens to make things worse and dampen our faith in our ways, and without it, we can finally move past this horrific moment in our fraught history."

The End

The Real Secret

Introduction: Setting the Stage

"Most self-help books are written by Americans of the most sentimental and unctuous sort. They promise their readers eternal life, untold riches and an escape from every grubby aspect of being human, all within 300 pages of upbeat, relentlessly repetitive and patronising prose. No wonder the unstated assumption of the cultural elite is that really only stupid people read them … What about everybody else?"

—Alain de Botton, British-Swiss philosopher

Self-help products and services are often treated as a joke in intellectual circles, but is that designation warranted? When you look at the current state of the self-improvement market, which is flooded with books filled with woo that are largely indiscernible from one another, I think it's easy to see how a reasonable consumer might say it is perfectly applicable.

In terms of books of advice that *don't* rely on debunked pseudoscientific principles, there isn't much. But that wasn't always the case. It turns out that self-help literature has a vibrant history of actually helping people. The self-help genre was originally created by socialists who encouraged readers to adopt a do-it-yourself attitude and acquire new life skills that might make it easier for them to be of service to others, especially within their local communities. That's according to Harvard professor Beth Blum, author of *The Self-Help Compulsion*, who speaks on the "radical" class-based origins of the self-help movement (Wilson, 2020). If that story is true, that's certainly a laudable goal. And whether the origin story is legitimate or not, it's clear that humans have always sought ways to become the best versions of themselves, and that religion and forced optimism were not necessarily part of that.

Jennifer Wilson, who writes for *The Nation*, *The New York Times*, and *The New York Review of Books* and won the 2022 Nona Balakian Award, gave some additional context.

"From the 1940s to the '60s, the Onitsha Market in southeastern Nigeria was the center of a burgeoning movement in self-help literature. Known as Onitsha Market literature, these cheap, locally published booklets were at once moralistic and titillating. With titles like *Why Harlots Hate Married Men and Love Bachelors*, *How to Avoid Corner Love and Win Good Love From Girls*, *Money Hard to Get but Easy to Spend*, and *Drunkards Believe Bar as Heaven*, they were sold at market stalls and were intended to appeal to a class of newly literate Nigerians interested in advancing themselves professionally and culturally," Wilson wrote. "Reading for self-improvement was something the British had encouraged in their mission schools, but the advice and moral guidance of the Onitsha Market pamphlets had much in common with the tenets of local folklore traditions."[1]

---

1. Jennifer Wilson, "Useful Books," *The Nation*, November 17, 2020, https://www.thenation.com/article/culture/self-help-compulsion-beth-blum-review/.

However, self-help began as a genre, it is clear that it has evolved greatly over time, as is the case with basically everything else on this Earth. Unfortunately, that change over time could more accurately be described as a *devolution*, considering the fact that most books that are lumped within that group now pander to the lowest common denominator and do virtually nothing to genuinely improve the lives of the readers. These books should *not* be considered self-help, and could perhaps more accurately be called self-deception, since the content—when there *is* anything substantive in them—is often untrue and/or generally unhelpful. I'm reminded of the episode of *Family Guy* in which a frustrated Brian Griffin writes a self-help book called *Wish It. Want It. Do It.*, which contains an abundance of blank pages to let the reader do all the hard work. This fictional storyline in the cartoon shows exactly the type of pseudo-intellectual books that make up so much of the existing marketplace in the real world.

When I initially considered the concept of *this* book, that was the current that I was swimming against. Facing wave after wave of books that do nothing but further dilute the purpose of the genre itself, the market drowning in repetitive and unnecessary nonsense, I decided I wanted this work to go back to the roots of *genuine* self-improvement efforts, and create something that can potentially provide a life raft that could help people lift themselves up and not only improve their own lives, but perhaps benefit their communities, as well. To that end, I engage with ideas from a wide range of thinkers, writers, activists, scholars, and public figures in this book. Quotations are included because they illuminate the subject under discussion, not because I endorse every belief, statement, or action of the individuals quoted.

Today, when it comes to books for helping you improve your life, you have tens of thousands of options—and a lot of those can be helpful. But that large sea of options also includes works such as *The Secret*, which is based on the supernatural (and demonstrably false) notion of achievement through nothing more than thought, and *The Purpose Driven Life*, which is essentially a book of religious affirmations pretending to describe a divine plan. Another one is *The Power of Now: A Guide to Spiritual Enlightenment*, which one TIME writer described as "*awash in spiritual mumbo jumbo*" and "*unhelpful for those looking for practical advice*."[2] All three of these mega-successful books, of course, are at the top of consumers' lists and were endorsed by talk show host and pseudoscience extraordinaire Oprah Winfrey. And this is just the tip of the iceberg, since thousands of other authors have released works that piggyback on these popular books in an effort to recreate that success for themselves. Each one of those new works seems to bring less substance than the last, instead focusing on repackaging the same feel-good ideas that made millions of dollars for others.

What's The Secret?

The Secret is a good example of the substanceless feelgoodery others are emulating. It has sold more than 30 million copies, yet it is based on one simple principle: thoughts *are* action. This basic idea is then labeled as the "law" of attraction, in order to give

2. Andrea Sachs, "Channeling Ram Dass," *Time*, April 14, 2003, https://time.com/archive/6668582/channeling-ram-dass/.

it unearned credence in the minds of consumers who falsely believe the concept is a law of physics. Using this "law" language, the author is trying to steer readers away from the fact that what they are engaging in is good old-fashioned "magical thinking." But, contrary to popular belief, magical thinking isn't all about magic lamps, rain dances, and voodoo chants. In fact, according to the Encyclopedia Britannica, magical thinking is defined as the "*belief that one's ideas, thoughts, actions, words, or use of symbols can influence the course of events in the material world.*" (Vandenberg, 2016).[3] In other words, magical thinking comes into play *any* time we feel that our inner thoughts are influencing our reality. That is the exact definition of the so-called law of attraction.[4]

How this process of self-actualization supposedly "works" is a bit difficult to nail down, but that is by design. The "law" is actually just a lie, couched in ambiguous statements about "quantum mechanics." Rhonda Byrne, who wrote *The Secret* and created the incredibly popular documentary by that same name, gives a prime example of this tactic. "I never studied science or physics at school, and yet when I read complex books on quantum physics I understood them perfectly because I wanted to understand them," Byrne states in the book, implying that she was able to *will herself* to understand the material because she wanted to. "The study of quantum physics helped me to have a deeper understanding of the Secret, on an energetic level." (Byrne, 2006)[5]

---

3. Brian Vandenberg, "magical thinking," *Britannica*, https://www.britannica.com/science/magical-thinking.

4. "Whatever you choose to think will become your life experience." —Rhonda Byrne, author, *The Secret*

5. Rhonda Byrne, *The Secret* (Australia: Atria Books/Beyond Words

Not only is this theory of self-actualization wrong, but it's dangerous, as well. It fosters a practice of inaction that helps justify bringing a knife to a gun fight on many crucial issues. Is homelessness a problem in your area? Just think hard about it. Are you worried about government corruption? Make a vision board. These are not *real* solutions, yet many seem to be convinced that doing *nothing* is essentially a life hack. Of course, a believer in this form of self-actualization might argue that it goes hand-in-hand with actions. But without those vision boards, actions are helpful. Without actions, on the other hand, the board is just material for recycling. If the vision board *needs* action to work, it is virtually useless. That's not to mention that some experts have said that self-reassurance collages and mantras can actually be counterproductive when they encourage individuals to sit back and relax while blessings come to them (Morin, 2021).

That being said, rituals and reminders can indeed be helpful, if properly utilized. So if you enjoy having a vision board, I recommend creating an evidence-based board with an emphasis on achieving outcomes. For instance, instead of compiling a board with photos of things you want, hoping that your wishes will go into the universe before they ultimately come true, you could include photos of realistic goals that you are working toward every single day. Then you could check off items as you achieve them.

Some people may not be looking for answers to anything, and are instead looking for a light and fun distraction from their lives, and for them perhaps *The Secret* may actually be helpful. But others of us need something more concrete than that.

Publishing, 2006).

That's what I was thinking about when I covered *The Secret* in my book, *No Sacred Cows: Investigating Myths, Cults, and the Supernatural:*

The Secret is a popular self-help book that teaches measurably false ideas, including that positive thinking itself (without further action) can create life-changing results, like increased wealth, health, and happiness. The hypothesis **essentially states that thinking specific thoughts can cause those thoughts to come to fruition in the form of various material items**. The falsely named law of attraction takes advantage of our human urge to believe in something greater, along with our pattern recognition capabilities and tendency toward confirmation bias, to **convince believers that their "vision board" is bringing things into their life**. While it's great to maintain a positive attitude and thinking about goals may help keep them in mind and therefore could make achieving them easier, The Secret is essentially **wishful thinking dressed up in New Age clothing**.

When I wrote that in *No Sacred Cows*, I didn't really think much more about what *The Secret*'s ineffectiveness actually meant to the general public. If you think more broadly, what it means is that the most popular works in a genre meant to improve readers' personal sense of well-being and, in some cases, even mental health, are modeled after pseudoscientific notions that *don't work*. And not only that, but they are actually encouraging people to remain stagnant, avoiding doing the real work that's almost always necessary to better anyone's life.

None of that even touches on the fact that *The Secret* uses several quotes that are divorced from their historical context in order to make its points. In fact, one writer, Julia Rickert, researched *The Secret*'s use of a purported Ralph Waldo Emerson quote using a librarian from Andover-Harvard Theological Library at the Harvard Divinity School, Emerson's alma mater. Rickert found absolutely no uses by Emerson of the alleged quote about "the secret" (Rickert, 2007). On page 183 of *The Secret*, the following quote is attributed to Emerson:

"*The secret is the answer to all that has been, all that is, and all that will ever be.*"

The quote doesn't appear to have ever been said by Emerson, but perhaps its inclusion was an honest mistake. So, where does the quote come from? Rickert had this to say about that particular subject: "To be clear, what I've come to suspect—strongly—is that Rhonda Byrne may have made it up."[6]

All of these factors combine to create a work that, in the context of science-based answers to real-world problems, doesn't cut the mustard.

Other Popular Self-Help Works

Books like *The Purpose Driven Life* and *The Power of Now* are no better than *The Secret* in terms of finding verified ways to help their readers, because these works too use fluffy language and invocations of the supernatural to make their points. At times, they do this without *really* saying anything of concrete value, yet some people eat it up because of the promises their authors make. These promises are, of course, often unfulfilled.

In the case of *The Purpose Driven Life*, critical reviews come

---

6. Julia Rickert, "A Little Secret about The Secret," *Chicago Reader*, May 31, 2007, https://chicagoreader.com/news/a-little-secret-about-the-secret/.

even from Christians who would largely agree with the author's perspective on religion. Tim Challies, a Canadian theologian and pastor, said, "Though I do not doubt his faith or his intentions, I find that the book itself deviates from Reformed doctrine on many points."

"The book is based on a false premise that there is supernatural value to a 40-day study. The author says that 'whenever God wanted to prepare someone for his purposes he took 40 days.' This is simply not true," Challies wrote. "Though the 40 day time period is used quite often in Scripture, we should not be superstitious about it. There are many examples of God taking different amounts of time to prepare people. Having to force the book to a length of forty days leads to a lot of repetition, especially in the last four or five chapters."[7]

The Power of Now has also received its share of criticism. Spiritual writer and speaker Akilesh Ayyar, for instance, says the author's "blandness, his confusion of categories, his poor thinking and lazy analysis of important texts … it's all so boring. Like a copy of a copy of a copy, it is tired, mixed-up, and cliched."

"I'm practically put to sleep by The Power of Now, the book that, after Oprah catapulted him to Deepak-Chopra (I have problems with Deepak, too, which I'll get to another day) levels of stardom, sold millions of copies," Ayyar wrote in his review of the work.[8]

---

7. Tim Challies, book review of Rick Warren's *The Purpose Driven Life*, January 12, 2004, https://www.challies.com/book-reviews/book-review-rick-warrens-the-purpose-driven-life/.
8. Akilesh Ayyar, "The Problems with The Power of Now," Sifting to the Truth, December 5, 2017, https://www.siftingtothetruth.com/blog/2017/11/7/the-problems-with-the-power-of-now.

These are just a couple of prominent examples of what has become an avalanche of biblically inspired works crushing the entire self-help genre. In fact, there are quite literally a *countless* number of these books. On my own personal bookshelf alone, I have nearly a dozen of them, including *Divine Healing: God's Recipe for Life & Health*, *God's Promises For Your Every Need*, and more.

Some purported self-help books attempt to disguise the fact that they rely upon religion, but the inspiration becomes clear upon reading. An example of this type of work is *Your Best Life Now: 7 Steps to Living at Your Full Potential.* The title doesn't scream "religion!" but the author is megachurch pastor Joel Osteen, and some of its chapters include "God Has More in Store," "Let God Bring Justice into Your Life," "Trusting God's Timing," "Trusting God When Life Doesn't Make Sense," and "Showing God's Kindness and Mercy."

"Unfortunately, what you will receive is directly connected to how you believe and what you expect," Osteen, senior pastor of the largest congregation in the United States, Lakewood Church, argues in the book. "If you want God to do the extraordinary, then you must start believing Him for bigger things."[9]

There are innumerable books like these on the market, and they all present a slightly different take when it comes to interpreting the biblical scriptures and other supernatural concepts, but that doesn't mean there aren't consistencies. One similarity among nearly all of these Bible-inspired works purporting to be about self-improvement, as well as non-religious supernatural self-help books like *The Secret*, is *inaction.*

---

9. Joel Osteen, *Your Best Life Now: 7 Steps to Living at Your Full Potential* (New York: FaithWords, 2004).

Because these works are based on spiritual or paranormal conclusions, their positions on self-help often rely upon *leaving things up to a supposed higher power*. Whether it's *The Secret* and the supposed law of attraction or one of Osteen's books about God and the power of prayer, the *answers* provided are about giving up control and not taking accountability for your own path forward. To me, this represents the opposite of the ideal approach. Instead of pushing things off in hopes that they will be solved by something or someone else, it's more productive to focus on what we can do to improve the situation for ourselves.

This issue isn't just found within self-help books; it's also present in religious texts themselves.

The Source Material

Let's step beyond actual self-help books for a moment and look at what *inspires* many of these works: so-called Holy Books. The Bible, for instance, is the inspiration for countless books claiming to make readers' lives better. To name just a few, we have *The Power of Positive Thinking* by Dr. Norman Vincent Peale, *Healing After Loss: Daily Meditations for Working Through Grief* by Martha Whitmore Hickman, *Resilient: Restoring Your Weary Soul in These Turbulent Times* by John Eldredge, and *Resilient Hope: 100 Devotions for Building Endurance in an Unpredictable World* by Christine Caine. These are all top-rated works on Amazon at the time of writing, and they all use Christianity and the Holy Bible as the foundation of their efforts to provide self-help. These types of books are easier to market, considering Christianity is the most popular religion in the world, with Christians making up more than half of the population of the U.S., but they don't necessarily provide real-world help rooted in evidence and experience. In fact, in many cases, they are devoid of original content and instead simply piggyback on out-of-context quotes and lessons mined from the Bible.

For an example of this self-help based on the Bible, we don't even need to read a book. Instead, consider Rick Warren, the Southern Baptist evangelical Christian pastor who wrote *The Purpose Driven Life*, founded a megachurch, and co-created "The Daniel Plan," a diet-planning system that purports to take its lessons from the Bible. Specifically, the system and accompanying book claim to be based on the prophet Daniel, who ate only vegetables instead of eating the foods provided to him by the king's court after Daniel had been kidnapped. Warren has been criticized for choosing this story as the basis for his diet system, in part because the story is *actually* about obedience, faith, and committing to following God's rules despite temptations. The actual *diet* has very little to do with the point of the story, yet Warren was able to launch an incredibly successful program based on that misinformation (Goodman, 2012).[10]

The Bible as a Self-Help Tool

Many Christians would argue that the Bible itself is the *original* self-help book. Barry Fredericks, pastor of Grace Christian Fellowship in Newtown, Connecticut, made that argument in an

10. Rob Goodman, "The Bible Is Not A Diet Plan," *Religion Dispatches*, August 14, 2012, https://religiondispatches.org/the-bible-is-not-a-diet-plan/.

article in a local newspaper. In that piece, Fredericks said, "As I read the Bible, it amazes me that so many principles that we live by and work by come right out of God's word."

"Corporate America spends billions of dollars each year on books and seminars that teach success principles. Virtually every success principle taught in these books and seminars comes right out of the Bible," he wrote. "The importance of character and a proper attitude are always stressed to employees and potential leaders. The Scriptures instruct us on not only the necessity for a character of integrity and a pleasant attitude, but also how to obtain them and make them an integral part of your life."

Fredericks also claimed that, "Other topics taught in books and seminars on how to succeed—like hard work, honesty, dependability and skill in the work you do—are all there in the Bible to train us how to do the kind of job your employer is looking for."[11]

As a Religious Studies graduate, I understand that, in reality, the Bible was created as a mechanism for regulating an ancient society and establishing tribal bonds. But it's true that, in certain cases, some people do use it to make themselves better people or improve their own lives. I don't oppose anyone using anything to help achieve their goals, as long as it doesn't come at the expense of others, but the fact is that there are better sources of knowledge and inspiration than archaic texts written at a time when day-to-day challenges were entirely different than they are today. In other words, if the Bible is helping you be a good person, then more power to you. But who knows if you

11. "The Bible is the original self-help book," *The News Times*, July 16, 2005, https://www.newstimes.com/news/article/the-bible-is-the-original-self-help-book-51546.php.

could be more ethical or more successful using other means? That's where science steps in.

When Self-Help Goes Wrong

Yes, self-help works are often associated with empty platitudes and even faulty advice. But in some extreme cases, the situation is much worse.

Self-help programs have historically been a hotbed for scams and abuse, in part because of an incredibly simple fact: those seeking to improve themselves and their lives are often among the most desperate and vulnerable. The entire religion of Scientology, which has been described as a "ruthless global scam" by investigative journalist Richard Behar,[12] is founded upon a form of self-help psychology popularized by writer L. Ron Hubbard. People looking to satisfy a curiosity or fulfill a spiritual need get sucked in by Scientology because of their hopes of a better life, and before they know it, they are paying the church thousands or even *hundreds of thousands* of dollars trying to pursue the "Bridge to Total Freedom."

It turns out that there is a pretty common theme, both in modern times and throughout history, of people seeking self-betterment becoming bamboozled. There are, of course, completely innocent and even helpful self-help ventures, as well, so individuals' experiences run the gamut. On the dark, terrifying end of that spectrum, one example stands out in my mind

12. Richard Behar, "The Thriving Cult of Greed and Power," *Time*, May 6, 1991, https://www.cs.cmu.edu/~dst/Fishman/time-behar.html.

far above the rest. I'm referring to NXIVM, the sex cult that was posing as a self-help/networking program to help users achieve business success.

In case you're not aware of the details, the scheme was the brainchild of a man by the name of Keith Raniere. Raniere had a background in pyramid schemes, having launched one that was shut down in the past. He then started NXIVM, which was also called "the vow," and was one part pyramid scheme, one part self-help workshop, and one part sex cult. The women who entered the program thought they would be exposed to a comprehensive plan to make them more successful in their careers, as well as to the already successful people who could help them along that path. In reality, they were forced to give up "collateral," AKA embarrassing materials, so that Raniere and others at the top of the sex-cult pyramid could blackmail them if they ever tried to leave the group.

This model isn't all that unusual in the world of cults. The fact is that there is an entire subcategory of cults that are focused on self-improvement, at least in name. One of the most successful of these cults was discussed above: Scientology. Scientology succeeded in using self-help as a mechanism to propel L. Ron Hubbard to cult leader status, and, since his death, the cult has continued to grow such that it is now often categorized as a religion and given similar protections as Christianity or Islam. Scientology has done all of this, while retaining its veneer as something meant to help people rid themselves of their past traumas.

Other self-help influences on cults are more subtle. Notorious criminal and cult leader Charles Manson, for instance, was heavily influenced by one particular self-help author: Dale Carnegie. It is documented that Manson turned to the works of Carnegie, the author of How to Win Friends and Influence People, to learn how to gain followers and control them. As Bloomberg reports, "Dale Carnegie Training, the self-help program that's shaped the lives of such people as Warren Buffett, Johnny Cash, and Emeril Lagasse, can claim an additional ardent disciple: Charles Manson."

"In his new book, *Manson: The Life and Times of Charles Manson*, author Jeff Guinn credits Carnegie Training with transforming Manson from 'a low-level pimp' to the 'frighteningly effective sociopath' who created a cult of killers in the late 1960s," the outlet reported. "Manson took classes in 'How to Win Friends and Influence People,' based on Carnegie's iconic book, while doing time for car theft in a California federal prison in 1957." Guinn said the program "was critical in shaping how he manipulated people."[13]

One of the worst scenarios like this, in which a purported self-help experience went wrong, involves *The Secret*. Specifically, the situation involved James Ray, a self-help guru who was known for having been interviewed in the documentary called *The Secret*. Seen in numerous videos advocating for the so-called law of attraction, Ray charged exorbitant amounts of money for his "Spiritual Warrior" sweat lodge experiences. In 2009, what likely started as nothing more than a scam ended in a deadly Arizona tragedy when three people died in Ray's sweat lodge. Jason Jones, an anti-cult activist and attorney, said that the way Ray did these events, twice as long as traditional and without

---

13. Diane Brady, "Charles Manson's Turning Point: Dale Carnegie Classes," *Bloomberg*, July 22, 2013, https://www.bloomberg.com/news/articles/2013-07-22/charles-mansons-turning-point-dale-carnegie-classes?embedded-checkout=true.

safety measures, was "inherently dangerous."

"It got more extreme even though people had such horrifying experiences in previous years," according to Jones, who called Ray a "cult leader."

Ray was ultimately convicted of three counts of negligent homicide, and he spent two years in jail for those crimes.[14]

Self-help programs have been at the heart of some terrible acts, but they can also be used for good things. It all depends on which course of *action* one takes.

---

14. Richard Siklos, "Remembering the Star-Crossed Life of Self-Help Guru James Arthur Ray," *Fortune*, January 17, 2025, https://fortune.com/2025/01/17/james-arthur-ray-obituary-self-help-guru/.

1 Action vs. Inaction

"The beginning is the most important part of the work."
—Plato

Many people have spent the entirety of their lives searching for the answer to one simple, yet also complex question: what's the point?

In other words, "*What is the meaning of life?*" Yet another variation of this question is, "*Why are we here?*"

These questions have occupied an incalculable number of hours for philosophers throughout all of time, but the fact is, there are multiple answers to each of them, depending on who you are and what your values represent. For some, the point of life is merely to procreate—to spread genes from one generation to another. For them, evolution shapes their purpose, which is genetically predetermined. For others, however, meaning can be found merely in the enjoyment of this life. Still others have a fixed purpose dependent on religion, another person, or just about anything else. In fact, there are probably as many "meanings of life" as there are individuals on Earth. It's just common sense that what some might consider to be meaningful could be largely irrelevant to others, and vice versa. It's a question that absolutely *must* be answered on a case-by-case basis.

The point of this book is not to propose an alternative view of "meaning," or to help you find some purpose that you lack. It isn't to show you that your passions are all wrong, and I do not intend to replace your interests with data or even science. Instead, I hope to present you with the best method for realizing your passion (or anything else). Of course, by that I mean simply using reason, combined with rigorous analysis of facts.

It seems like common sense, doesn't it? Of course, a scientific view, combined with sharp critical thinking, is the best way to achieve practically anything. But that's not what most people jump to when they need to accomplish their goals. In fact, it seems like most people—at least in the United States of America and the West more generally—take a completely opposite approach.

A Persistent Problem

Instead of working harder, following the facts, and using reason, many people resort to prayer, positive thoughts, and reliance on mysterious relics combined with some form of detachment from the situation itself. In the U.S., this *active inaction* tactic is perhaps most well-known when it comes to the mass shootings debate. Every time a mass shooter loses it in a traumatic attack that takes the lives of innocents, some people try to will the problem away with religious rhetoric, while others seek to address the underlying issues. This isn't a political concern, either. One doesn't have to have a position on this topic (or on gun control generally) to understand that the offering of "thoughts and prayers" in the wake of such a tragedy is *literally the least you can do*. It offers the pray*er* (person praying) a sense of satisfaction, while the pray*ee* gets nothing. It is working as a coping mechanism for those on the outside looking in, without ever addressing the real problem. Whatever you believe the solutions are to this complex problem, I think we can all agree that prayer has not worked to solve it.

But it's not just about the debate over the power of prayer versus concrete actions. It's pretty clear that this same way of thinking permeates nearly every corner of our society. There are bestselling books that not only encourage thoughts over actions, but paint the mentality as the answer to all life's problems, and blockbuster movies all about how it is best to sit back and wait for blessings to come to you. People will go see mediums or influencers instead of real doctors for medical problems. Purportedly wise individuals are quick to hit you with clichés like, "*Good things come to those who wait*," and "*It's all in God's hands now*," even when doing so directly discourages actions one could take to address their concerns. It's important to note there is nothing inherently wrong with these sayings, as of course, patience is itself a virtue, but "patience" at the expense of reasonable action is just plain old inaction. And that is far too common these days, and can be harmful when solutions *do* exist.

Of course, I am no psychic. But I predict many people reading this are starting to get defensive at this point. You might be thinking, "I would *never* offer a prayer when I could help someone in need," or, "This person is crazy if they think I would see a 'guru' to solve my problems." But, even if you don't use prayer as a substitute for action, and don't read books that promote not doing anything or listen to mystics who just want your money, it's still possible that you are choosing passive paths toward your goals over more active (and infinitely more useful) ones. If you've ever posted a status update on social media to promote a certain law, instead of calling your local law*maker*, you've experienced this yourself. [1]

Social Slacktivism

One trap that well-intentioned individuals can fall into when trying to make a difference is the online activism black hole. Of course, there have been incredibly valuable achievements in the world of activism via social media—for instance, large-scale demonstrations that were organized online and then moved into real action, or even online protests that resulted in concrete government reform—but there is also a darker side to this method of solving problems in the world. I'm talking, of course, about the fraudulent feeling of accomplishment that can come along with certain forms of regressive social activism. It's the same feeling that can cause the so-called law of attraction to backfire, i.e., when someone needs a new job but feels as if their vision board is doing the trick, so they don't feel like they need to apply to as many positions as they otherwise would.

Although I have at times been critical of social slacktivism, and am doing so again here, it's also important to note that I have also been guilty of it in certain cases. I have shared something to social media and thought to myself, "Oh, good, I'm doing something helpful." Did that keep me from donating or

---

1. This will be discussed further in the second chapter, "Doing the Work."

even volunteering separately? Of course not. Not literally. But I can't deny the *possibility* that there have been times when my feeling of accomplishment overcame my desire to do more, and that somewhere out there, someone was worse off because of that inaction.

I'm not the only one. Whether we like to admit it or not, we have all done this. We have all felt content with what we've done after participating in some form of online or otherwise abstract activity, when we could actually be doing more, easily. But that doesn't mean you can't choose to do more. We can always choose to *take action*.

Considering the Evidence

So, aside from simply *taking action*, what are we talking about here? The next ingredient is *evidence*. To see why evidence is important, imagine if you get a call from a friend, saying their house is on fire and they need your help. They say not to ask questions, and they ask you to bring water to them right now. You grab some buckets, fill them with water, and head over to your friend's house. But when you get there, you see there was never any fire. Instead, the building appears to be flooded. And it's confusing. You did *everything* right, yet you ended up in a bad situation because you had some bad information. To make matters worse, you got the faulty information from a source you trusted! And this happens all the time in your daily life, even if it is on a smaller scale.

My point is that action is incredibly important, but you can *act* all you want, and those actions are meaningless if you are acting on bad data. If you knew your friend's house was flooded instead of on fire, perhaps you could have brought a Shop-Vac or maybe a hand pump, as opposed to buckets of unnecessary water that would make things worse. So, getting the data right is actually the most important aspect of making decisions.

But data itself will never tell us all the answers. If decision-making is painting, the action is the motivation, and the data are like the paints. But that's not all you need to make a master work of art: you need brush strokes.

That's why it's important to analyze data, factor in peer-reviewed scientific studies, and then make thoughtful decisions and actions from there.

The Brush Strokes

Unfortunately, data will never be enough to properly inform every single decision you face, such that you only see positive outcomes (although some strict empiricists would argue it is). That's where *reason* comes in. It's not just a buzzword for politicians looking to deceive; it also means using critical thinking to determine what the most reasonable course of action is. This can include thought experiments, philosophical and ethical touchstones, and much, much more.

When you combine these different viewpoints and methods of addressing various problems, you discover that it's not all that difficult to handle issues more rationally. Humans tend to take mental shortcuts that keep us from being logical in every situation, but it is entirely possible to overcome that. In other words, there is a *reasonable* way to use all the information available and make legitimate, lasting improvements in your own life.

In my earlier works, I've written about how to use science and critical thinking to your own benefit. Specifically, I've discussed how to use these processes to decide which products are best for your family and uncover which advertisements are actually being untruthful. What I *didn't* talk about is how you can also use these same mechanisms to accomplish household chores in a more efficient manner, or how much money and time you would save if you treated every day like it was April Fool's Day. This is a day known for practical jokes, but it's also a time when people pay close attention to the happenings around them in an effort to avoid being *fooled*. It could instead be called International Day of Skepticism, because it's the only day during which many people actually question what they see, and research whether or not things are true (if they happen to remember the date). As a result, some people get tricked *less* often on this day. And this is a mentality we can all carry on into our everyday lives merely by asking questions and paying more attention to what's happening around us.

Science of Spirituality

Achieving more and being more logical about day-to-day choices is certainly important, but is that all life is to you? For most, there's more. For instance, inspiration and spirituality are important aspects of many people's lives, yet science and reason are largely left out of those conversations. Is that warranted? According to the legendary astronomer Carl Sagan, the simple answer is, "No." As he wrote in *The Demon-Haunted World*, "*Sci-*

ence is not only compatible with spirituality; it is a ***profound source of spirituality****. When we recognize our place in an immensity of light years and in the passage of ages, when we grasp the intricacy, beauty and subtlety of life, then* ***that soaring feeling, that sense of elation and humility combined, is surely spiritual****."*

The importance of this element of so-called scientific spirituality can't be overstated, because the curiosity and passion that drives science forward can also be used to harness some of the most amazing feelings. Science isn't just a game of cold data, as some people believe, and with this book, I hope to walk people through how to learn that firsthand.

It's established that many people want to feel better and do better, and in an era when people are fed up with the "Facts don't matter" mentality, I think they don't want to settle for false promises made by the most popular spiritual books. I believe most people, given the choice, will choose something that will help them succeed more in a way that is based on solid, verifiable data. And that's exactly what I hope to provide.

We use science for so many things in this world, so why can't improving ourselves be one of them? Surely, it can.

2 The Real "Secret" Is Hard Work

"The dictionary is the only place that success comes before work. Work is the key to success, and hard work can help you accomplish anything."

—Vince Lombardi

Let's start off with a simple exercise to get you to open up. I want you to clear your mind entirely and just breathe. Then picture a large piece of cardboard. It's blank and sitting in front of you. Now, imagine yourself placing images on the board. Images of things you love and want. Things you hope to be able to afford one day, or have secretly hoped for. It could be anything—a new car, a new home, a new career—*anything*. Now picture yourself throwing that board in the trash, and then snap back into reality. Vision boards, like the so-called law of attraction, don't help you achieve your life goals. The grain of truth in these pseudoscientific practices is that they *can* help you work toward your goals if you are constantly thinking about them, but the key component is still *work*. And it's also true that, if you believe you have accomplished something just by visualizing it, you can actually hamper your own efforts to improve your life.

There will always be popular works that promise magical loopholes and secret-to-success shortcuts meant to get you rich quick. This is *not* one of those works. While affirmations and positive thinking are great, and certainly have their place in any healthy individual's daily life, they do not typically *solve* our problems. In fact, some studies have shown that practices encouraged by some spirituality-based self-help books can cause people to *feel* as if they've accomplished something, leading them away from doing any further (actual) action. Of course, we all know at least one individual who swears that hoping for something is equal to taking significant steps. They could be experiencing a placebo effect, or maybe they're just lucky. Or it's possible they are seeing effects of the Baader-Meinhof phenomenon, or frequency illusion, which Psychology Today defines as "*a cognitive bias in which someone learns a novel word or concept—and then 'suddenly' encounters it everywhere, whereas in fact it is just more salient because it has been recently observed.*"[1]

Here are some examples of the phenomenon brought to you by the University of Colorado, Denver: "*Let's say you learn that Asmara is the capital of Eritrea. A week ago, you didn't even know there was an African country named Eritrea. Now every Uber driver you meet happens to be from … you guessed it, Eritrea. This is a strange phenomenon related to memory—and it's called the Frequency illusion. Now that you know about X, is X starting to show up everywhere or was X always there and you've just become attuned to noticing it? That is the million-dollar question, as it turns out.*"[2]

---

1. "Frequency Illusion Baader-Meinhof Phenomenon," *Psychology Today*, https://www.psychologytoday.com/us/basics/frequency-illusion.
2. Alicita Rodriguez, "What Is the Frequency Illusion," *CU Denver News*, September 24, 2019, https://news.ucdenver.edu/what-is-the-frequen-

Regardless of how we feel, the fact is that there is no "secret" answer to making life better. There is no *secret* because the method that can actually help improve your life is one that is quite battle-tested: you have to work hard. This is the difficult truth that not everyone is able to accept, and the reason so many people turn to self-help systems that emphasize religion or some other supernatural force. It's not easy to accept that, sometimes, putting your nose to the grindstone is the only way to get results, no matter what your life coach tells you.

But once you've accepted that fact, there are next steps. And how you choose to go about your work will be a major factor in whether or not it's successful. Once again, it will sound crazy to some, but there is an answer: I'm talking about the controversial practice of using science, reason, and compassion to guide your daily behavior and achieve goals. While there is no *perfect* way to approach life, and no one tactic that can be applied to every single situation, evidence suggests that using these three ideas together can yield results that rival all else. Accepting the realities about hard work and getting to real-world solutions also means understanding that the important answers are rarely the easiest ones. I'm aware that this solution of looking for scientific answers isn't appealing to everyone, in part because it isn't a clear-cut, one-size-fits-all approach. But that is just how the world works. I know it would be easier if it were, but life isn't black and white. Almost nothing is. Some of the biggest controversies in the world could be solved if we stopped looking for one answer and searched for a set of solutions depending on varying scenarios. Therefore, taking things on a case-by-case

---

cy-illusion/.

basis helps you achieve better results.

While it's clear what methods help people, it's not always obvious to everyone how best to apply them. That's where "wisdom" from self-proclaimed self-help gurus often comes in. But by positing that *thought* itself is an action, which has become an all-too-common refrain of the spiritual self-help crowd, those purported leaders are actually drawing attention away from the fact that action is *always* the best way to get real goals accomplished. As poet and independent rapper George Watsky has said, "Waiting doesn't work, and ***praying may not come through***, and hoping doesn't work, **so I will be the one to ... *work***."

Types of Inaction

Traditional forms of inaction include inaction from *ignorance*, which comes into play when a person simply isn't aware that they need to do something, and inaction from malice. That could mean hatred if you aren't acting to help someone else, or, in this case, it could be self-sabotage if you refuse to help yourself.

Let's look at this in a scenario:

Imagine you work in an office, and you have been eyeing a promotion for months. You *feel* you deserve it, and it would really help your finances, but it is *not* going to just come to you. If you read *The Secret* and believe it to be based in reality, you might go home and draw a picture of your promotion. Maybe you'd draw yourself shaking hands with your boss, and viewing the image might give you a temporary boost of feeling accomplished. But what *did* you really accomplish? Nothing.

Given this same scenario, let's say you look to religious scripture instead of self-help books. You might come out of your Bible (or Qur'an, or basically any Holy Book) study thinking that God has a plan for you, and that He will give you your coveted promotion. You might go to bed with a smile on your face, believing a higher power is in charge, but in reality, you are exactly zero steps closer to actually getting what you wanted.

The only *real* solution won't be found in these books, perhaps because it's too simple: you should work hard, communicate your goals to your superiors, and show your employer what you can accomplish. But because this isn't a *shortcut*, people don't want to hear it.

If you're feeling good about yourself right now, like you're above this type of cultural inaction because you don't rely on higher powers to act on your behalf, let me explain exactly why you're wrong.

The fact is that we all engage in inaction when we should be *doing something*. This could come in the most benign forms, such as mere procrastination, or it could be a more persistent issue, like *slacktivism*, which we discussed in the previous chapter. It's also worth noting that, when I say slacktivism, I am referring to any type of action that makes you feel like you're doing something when you're not moving the needle. A common type of slacktivism is *internet* slacktivism, which usually takes the form of signing petitions and sharing posts as a means to feel like one is solving a particular problem.

Internal Impacts

Slacktivism is a big issue on a societal level because it keeps worthy acts from being accomplished. But it can also affect individuals, especially as it relates to how accomplishing tasks—and doing *good* things more specifically—makes us feel.

Science has shown that we tend to focus on completing mundane, everyday tasks, as opposed to working on those long-term goals that might bring us closer to success in life. But why is that the case? According to Harvard Business Review, this is in part because of the way our brains are wired when it comes to accomplishing things. When we complete something off our to-do list, for instance, our brain releases dopamine.

"One of the main reasons this happens is that human brains are wired to seek completion and the pleasure it brings—a tendency we term 'completion bias.' Completing simple tasks, such as answering emails or posting updates on your Twitter account, takes little time and allows you to check off items on your to-do list. Our ongoing research has found that checking off items is psychologically rewarding: After you complete a task, being able to literally check a box makes you happier than when you are not given a box to check," write Francesca Gino and Bradley R. Staats. "Complicating matters, finishing immediate, mundane tasks actually improves your ability to tackle tougher, important things. Your brain releases dopamine when you achieve goals. And since dopamine improves attention, memory, and motivation, even achieving a small goal can result in a positive feedback loop that makes you more motivated to work harder going forward."[3]

3. Francesca Gino and Bradley R. Staats, "Your Desire to Get Things Done Can Undermine Your Effectiveness," *Harvard Board Review*, March 22, 2016, https://hbr.org/2016/03/your-desire-to-get-things-done-can-undermine-your-effectiveness.

To put these findings in practical terms, two things are clear:

1. Because even minor goal completions give similar chemical brain boosts as major successes, it's possible to become distracted with smaller things.
2. Those brain boosts—small or not—can potentially help create a positive feedback loop of completion.

Given these two factors, it makes sense that, to achieve your best efficiency, you could dedicate some time to getting a number of things done, and then begin with the easiest ones to help get the ball rolling, before moving to major tasks in a structured way.

International Correspondence Schools Limited, a global online learning organization, goes further in describing how chemicals in our brain relate to making goals.

"Dopamine and serotonin in particular are the neurotransmitters that are relevant when it comes to goals. Dopamine is considered a 'reward' chemical that your brain releases when you've achieved (or you're close to achieving) a goal. It creates feelings of motivation, satisfaction and productivity. Serotonin helps to create feelings of happiness, calm and focus. The presence of neurotransmitters explains why you feel great when you achieve a goal," writes Christopher Fear. "Synapses—the points where neurons communicate with each other—are intimately tied up with how the brain sets goals, and understanding how can help you improve your chances of success."[4]

4. Christopher Fear, "The Science of Setting Goals—and How to Achieve Them," ICS Learn, March 17, 2022, https://www.icslearn.co.uk/blog/study-advice/the-science-of-setting-goals-and-how-to-

Learning about and mastering this connection between our goals and our brains is one key to ensuring you get the best chance at succeeding in nearly anything.

Our brain chemistry also comes into play when it comes to doing specifically *good* things. According to a blog by Cedars-Sinai, "The warm feeling of wellbeing that washes over you when you've done something kind isn't just in your head."

"It's in your brain chemicals, too," according to the blog. "Acts of kindness can release hormones that contribute to your mood and overall wellbeing. The practice is so effective it's being formally incorporated into some types of psychotherapy."

Cedars-Sinai quotes Dr. Waguih William IsHak, a professor of psychiatry at the hospital, as saying, "We all seek a path to happiness … Practicing kindness toward others is one we know works."

"The trick you need to know: Acts of kindness have to be repeated," Dr. IsHak added. "Biochemically, you can't live on the 3-to-4-minute oxytocin boost that comes from a single act."[5]

A Vision Board That Works

Using a "vision board" to "manifest your deepest desires" is thought by many people to be a type of magic/magick, but others believe it to be a science. Both assessments are incorrect. In reality, this is more of a pseudoscience, with the added benefit of a placebo effect.

---

achieve-them.

5. "The Science of Kindness," Cedars-Sinai Blog, February 13, 2019, https://www.cedars-sinai.org/blog/science-of-kindness.html.

The notion is that you place visual representations of your desires in a prominent place, such as adhered to a piece of school project cardboard, and look at the images every day. By the so-called law of attraction, which is not a real thing, it is believed that these items will be brought into your life because of your thinking about them. The energy of your thoughts goes out into the world and "attracts" the subjects of those thoughts, according to believers.

Now, this is all nonsense. There is absolutely zero evidence that the "law of attraction" is even real. What *is* legitimate, however, is the idea that thinking about your goals every day can be beneficial. This is the nugget of truth that has caused ideas like the law of attraction to thrive. But it's not because your thoughts have power and bring your desires closer to you—it's due to the fact that thinking about your goals will put them top of mind, making your daily actions more in line with them. For example, if you are constantly thinking about how you need a promotion, you might make the small choices that get noticed by your peers and your leadership. Similarly, if you are thinking non-stop about how you need a new car, there's a chance you'd be more likely to put aside the money needed for your down payment, or perhaps that you make the needed improvements to your credit to make such a major purchase possible.

All that being said, it's possible to use a vision board that *does* use scientific principles to help you improve your life. You would simply follow the same steps as a typical vision board—choose a specific goal or objective to focus on, find images, quotes, or other visual elements that represent your goals, arrange the images on a board or poster, and display it somewhere where you can see it on a daily basis—the difference would be in the execution. Are you using the board to make yourself feel better about hopefully one day achieving your goals, despite doing nothing tangible in your everyday life to bring them into your reality, or do you use it as a reminder so that you remember to match your daily actions to your long-term goals, thereby achieving success?

3 Separation of Church and Self-Improvement

"Secular people don't believe in life after death, but rather, they believe in life before death."

—Phil Zuckerman, *Living the Secular Life*

Religion is at the core of nearly all self-improvement books, videos, programs, and organizations. In fact, in the U.S. justice system, people who are convicted of minor drug or alcohol-related crimes are often ordered to attend theism-based organizations as part of their sentencing, further demonstrating the firm links between spirituality and our society's idea of self-care/rehabilitation.

While this situation is not ideal, it's perfectly *understandable*, if you think about the context in which religion is connected to society. After all, "religion" might be considered one of humanity's first attempts at self-improvement or at least ethical regulation. Even today, certain religious belief systems are at times used to help people turn their lives around. And if that works for them, I'm all for it!

But not everyone is compatible with religious ideologies that require, for instance, that the individual push their problems onto an unknown deity. For many people, improving their lives is about taking responsibility for their actions, holding themselves accountable, and charting their own way forward. These goals simply don't compute if a person who wants to work on themselves is forced to do so within a religious framework that emphasizes giving up their agency to a higher power that has already planned their entire life from beginning to end.

This very simple concept that religion shouldn't be forced on anyone was recognized by the founders of the United States, as represented by the wall they built between religion and the government of the country. The notion of a "*wall*" came directly from Thomas Jefferson himself. In a letter to a group of Danbury Baptists, Jefferson said, "*I contemplate with sovereign reverence that act of the whole American people which declared that their legislature should make no law respecting an establishment of religion, or prohibiting the free exercise thereof, thus building a wall of separation between church and state.*"

This is what I propose in connection with works that emphasize making one's life better. I don't want to stop anyone from believing anything, and, in fact, if religion is a part of someone's journey toward being happier and healthier mentally, I fully support that. But when it comes to the general public, which is made up of people from every walk of life, it might not be necessary to force the square peg of religion into the round hole of self-help. A lot of people are interested in secular, science-based mechanisms for making things better, as opposed to lofty aspirational quotes based on an individual author's interpretation of sacred scriptures.

There's no doubt that religion can at times be a positive

force for the individual. In some cases, religion has been used to get people off drugs or to encourage charity for the less fortunate. But there's also nothing that suggests religion should be inserted into every area of life, or that these types of positive acts can't be achieved without a dogma.

So, if we are to build a wall of separation between Church and Self-improvement, where should we begin its construction?

Christianity didn't invent charity. Stop pretending it did.

For me, one of the things that helps me feel better in my life is helping others who need it most. But I've noticed a common false belief linking these types of actions to religion or, specifically, Christianity. In fact, one of the most common arguments I heard from Christian students in my Religious Studies program was that Christianity must be true because of all the charity work Christian churches do. This is not only a false connection, but it's also a misrepresentation of reality.

I'm not one to deny facts, and I must give credit where credit is due. It's true that Christians are responsible for a substantial number of charity dollars given each year, especially if you include tithes to a church as charitable giving, which nearly all surveys do, despite a lack of transparent accounting as to where that money really goes.

For instance, the Pew Research Center has found that "*volunteerism and donations to the poor are especially common practices for those who are highly religious.*"

"Among people who pray daily and attend services weekly, 45% also say they volunteered in the past week (**including 23% who did so mainly through a church or other religious organization**)," the center reported. "Just 28% of Americans who are not highly religious say they volunteered in the past seven days. The gap is even bigger when it comes to helping the poor: 65% of the highly religious say they donated money, time or goods to help the poor in the past week, compared with 41% of all other U.S. adults."[1]

Carefully woven within this data is the fact that church events and tithing are both considered charity. When you subtract the 23% who did volunteer work mainly through their church, the percentage among highly religious people is closer to 22%, compared to 28% of "*not highly religious*" respondents who said they did volunteer work within the last week.

There's also a false piece of information floating out there somewhere that secular and atheist groups and societies don't focus on charity and helping others. In reality, social improvement is at the heart of many secular and skeptic groups' goals. American Atheists' National Atheist Convention, for instance, places a strong emphasis on helping others.

"The festivities kick off on Thursday evening with a Charity Game Night benefitting Access Reproductive Care—Southeast, an organization working to ensure that Southerners receive safe and compassionate reproductive care including abortion services," the group wrote in a past press release. "On Easter Sunday, atheists will come together for the 'Two Hands Working' Service Project. This volunteer project is named after Madalyn Murray O'Hair's famous quote, 'Two hands working can do more than a thousand clasped in prayer.' During the event, attendees will pack 50,000 meals for people in need in the Atlanta

1. Michael Lipka, "How Highly Religious Americans' Lives Are Different from Others," Pew Research Center, April 12, 2016, https://www.pewresearch.org/short-reads/2016/04/12/how-highly-religious-americans-lives-are-different-from-others/.

area."[2]

I've personally volunteered with Atheists United, which is known for sponsoring trash clean-ups in Los Angeles.[3] And I can also say with certainty that, when I was homeless, I remember Christians and atheists alike volunteering to help me.[4]

I've also heard the argument that Christianity *invented* charity work, which is just historically false. Not only was charity prominently featured in Judaism (tzedakah), on which Christianity is based, but examples go even further than that. There's also the concept of dāna, which is present in Hinduism, Buddhism, and Jainism, all of which predate Christianity.

All of that taken into account, it's worth noting that atheists and believers alike do tend to give regularly. However, if you are *only* giving to a charity because you believe a deity told you to, is it really coming from the right place?

Recovery Programs (AA and NA)

A prime example of the need to separate religion from self-improvement, popular drug and alcohol recovery groups are almost always built upon religion or, at very least, a vague form of

---

2. "National Atheist Convention in Atlanta Feeds the Hungry and Funds Abortion Access on Easter Weekend," American Atheists, March 31, 2022, https://www.atheists.org/2022/03/american-atheist-convention-atlanta-hungry-abortion/.
3. Volunteer to clean up trash in Los Angeles with Atheists United: atheistsunited.org/highway-cleanups.
4. David G. McAfee, "I Was Homeless as a Kid, and Here's What It Was Like," Patreon, July 31, 2022, https://www.patreon.com/posts/i-was-homeless-69819904.

theism that I would argue is counter-productive when seeking real help for addiction issues.

Most of these recovery groups are based on a 12-step program developed under the Alcoholics Anonymous banner, although Narcotics Anonymous is another large, related group, and there are several others.

As a kid, it was actually in the rooms of N.A. that I had some of my first introductions to religion, God, and the notion of passing the buck when it comes to responsibility for one's actions. With both of my parents being addicted to meth, I spent between one and three nights per week in N.A. meetings starting at seven years old. There were a *lot* of good people in those rooms, and I truly do think many of them were better off as a result of having that structure and community. But there's also no doubt that religion and supernatural belief were at the heart of the program itself, when it wasn't needed. I would even argue that, at times, N.A. served as a mini-religion of its own—or perhaps more accurately, a cult with its own unique rituals and hierarchies.

I heard hundreds, maybe thousands of testimonials in those rooms. And 99.99% of them had one thing in common: they used God and religion as a crutch. A lot of these individuals were helped by the program, but that doesn't change the fact that each and every one of them had given up control of things in their lives to something or someone else.

These were my experiences in N.A., but it is much bigger than that. Because these groups are based on Alcoholics Anonymous, it might help to get a brief overview of how A.A. and its "Big Book" came to be.

The book, A.A.'s Bible, for lack of a better term, was first published in 1939, according to the group's website. A.A. says its "*purpose was to show other alcoholics how the first 100 people of A.A. got sober*."[5] But in the book itself, the purpose is stated as such:

"Lack of power, that was our dilemma. We had to find a power by which we could live, and it had to be a Power greater than ourselves. Obviously. But where and how were we to find this Power?" the book asks. "Well, that's exactly what this book is about. Its main object is to enable you to find a Power greater than yourself which will solve your problem. That means we have written a book which we believe to be spiritual as well as moral. And it means, of course, that we are going to talk about God. Here difficulty arises with agnostics."[6]

And the Big Book lives up to this promise to be all about God. In fact, the word "God," or some variation of "Him," is reportedly found 281 times in the first 164 pages of the Big Book.[7]

A lot of A.A. followers will be quick to point out that, in practice, the people in the rooms of these recovery groups aren't likely to force religion on anyone. They actually put in a lot of effort to steer away from specific deities tied to today's faiths. Instead, they focus on part of what was quoted above: "a Power greater than ourselves."

For many 12-step members, this phrasing circumvents the idea of a Christian or Muslim or Hindu deity. Even atheists and

---

5. "About the Big Book," Alcoholics Anonymous, aa.org/the-big-book.
6. Bill Wilson, *Alcoholics Anonymous: The Big Book* (U.S.A.: Alcoholics Anonymous World Services, 1939).
7. "Why Is the Big Book Sacred?," AA Agnostica, https://aaagnostica.org/2020/04/12/why-is-the-big-book-sacred.

agnostics have been known to adopt this language of a "Power greater than ourselves" if they are participating in these groups. But for me, this doesn't solve the issue at all.

The biggest problem with reliance upon the supernatural for these programs, or self-help more generally, isn't tying the therapeutic efforts to a religion with a dogma. That's, of course, *a* problem, but even bigger than that is this notion that we must build ourselves up based on an abdication of responsibility.

For example, take a look at the first step in A.A., which states, "We admitted we were powerless over alcohol — that our lives had become unmanageable."[8] The second step, which builds off that, states, "We came to be aware that a Power greater than ourselves could restore us to sanity."[9]

This is the problem. To participate in these recovery groups, you have to acknowledge that you are "powerless" and that there is a "power" greater than yourself who will help you solve your problems. Under a science and reason-based approach, it's the opposite: you want to take complete responsibility in a way that enables you to take control of your own life and not rely on some mystical force to do it for you. This brings us back to the original question of actions vs. inaction and work vs. prayer.

A lot of A.A. adherents will still *insist* that their system isn't inherently religious. While I feel the opposite point has already been made, I will also refer to this quote from the U.S. Department of Justice's Office for Civil Rights in answer to the question, "Is a twelve-step recovery program such as AA or NA an

8. "Step 1 of Alcoholics Anonymous (AA)," American Addiction Centers, https://alcohol.org/alcoholics-anonymous/step-1/.
9. "Step 2 of Alcoholics Anonymous," American Addiction Centers, https://alcohol.org/alcoholics-anonymous/step-2/.

inherently religious activity?"

"Yes," the office affirms:

"Courts have analyzed this question in reference to the Establishment Clause of the First Amendment to the United States Constitution. The Establishment Clause prohibits the government from making any law 'respecting an establishment of religion.' This clause not only forbids the government from establishing an official religion, but also prohibits government actions that favor one religion over another or prefer religion over non-religion. Courts have repeatedly found that traditional twelve-step programs contain religious content and are religious activities. In reaching this determination, courts have relied upon several factors, such as the fact that the twelve steps are based on the monotheistic idea of a single God or Supreme Being, which is rooted in the religious concept of a Higher Power. It is important to keep in mind that a program or activity does not need to be a traditional form of religious worship to be a religious activity; a program can have a secular purpose such as providing drug or alcohol treatment but its programming may contain religious content."[10]

The 12-step system is a key form of self-help that is unnecessarily injected with religion, but it's far from the only example.[11]

---

10. "Frequently Asked Questions Regarding Twelve-Step Recovery Programs for Recipients of Justice Department Financial Assistance," U.S. Department of Justice: Office for Civil Rights, https://www.ojp.gov/sites/g/files/xyckuh241/files/media/document/twelvesteprecoveryprograms_faqs.pdf.

11. I cover this topic in *Hi, I'm an Atheist! What That Means and How to Talk about It with Others*. In that book, I write about the "people who claim to need religion to cope with the realities of day-to-day life—or to justify their morality": "In fact, for those individuals, a religion may

Religion in Prison

Contrary to popular belief, the purpose of the U.S. prison system is not only to *punish* individuals for the crimes they committed (nor is it to line the pockets of lobbyists and private prison operators, although that certainly happens). One reason our *corrections* system exists is that, in most cases, offenders need to improve themselves before reentering society. According to the U.S. Department of Justice, its rehabilitation programs "are designed to reduce recidivism among adults by improving their behaviors, skills, mental health, social functioning, and access to education and employment."[12] In some cases, these programs are alternatives to incarceration.

Religion has had its influences on U.S. confinement since its earliest stages. In fact, the very foundation of our prison system is built upon a system created by reformers mimicking certain monastic practices, according to experts.

"In devising this alternative," explained Andrew Skotnicki,

provide a sense of well-being in an otherwise overwhelming world, but it rarely leads people to solve their problems; it often only encourages them to leave these issues up to the mystical higher power and dodge responsibility. This has been seen, for instance, in some implementations of the 12-step recovery programs based on Alcoholics and Narcotics Anonymous. As part of that process, participants (even non-believers) are asked to accept that they can't change certain things and leave parts of their lives up to an undefined 'Higher Power.' It's easy to see how this mentality, if taken to its extremes, could lead a person toward inaction in their daily lives in lieu of petitionary prayer."

12. "Practice Profile: Rehabilitation Programs for Adults Convicted of a Crime," Crime Solutions, National Institute of Justice, February 11, 2020, https://crimesolutions.ojp.gov/ratedpractices/rehabilitation-programs-adults-convicted-crime.

a professor of religious studies at Manhattan College in New York City, "these early prison reformers drew on the monastic practice of confining a wayward monk to his cell, supplying him with work and letting time heal so that he might be returned to the community."[13]

Despite prisons and jails being meant in part to help reduce future crimes by improving the habits and mental health of prior offenders, in practice, this hardly seems to be the case. Yet the religious influences have grown stronger. Religion has many key roles in the modern confinement system in the U.S., and is even used as a justification for prisoners to get parole or early release. There's also an incredibly active chaplain system in the U.S., which allows religious figures to work one-on-one with, and even convert, convicted felons. More than seven-in-ten (73%) state prison chaplains say that efforts by inmates to proselytize or convert other inmates are either very common (31%) or somewhat common (43%), according to Pew Research Center.

"About three-quarters of the chaplains say that a lot (26%) or some (51%) religious switching occurs among inmates in the prisons where they work. Many chaplains report growth from religious switching in the numbers of Muslims and Protestant Christians, in particular," the center found in 2012. "Overwhelmingly, state prison chaplains consider religious counseling and other religion-based programming an important aspect of rehabilitating prisoners."[14]

---

13. Betsy Shirley, "Religious Ideals Shaped the Broken U.S. Prison System. Can They Also Fix It?" *America: The Jesuit Review*, July 17, 2020, https://www.americamagazine.org/politics-society/2020/07/17/religious-ideals-shaped-broken-us-prison-system-can-they-also-fix-it.
14. "Religion in Prisons—A 50-State Survey of Prison Chaplains,"

All of this considered, it might not be so surprising to learn that the levels of religiosity in prisons are much higher than in non-confinement settings. According to one British journalist, "Overall, almost 1 in every 1,000 prisoners will identify as atheist compared to 1 in every 100 Americans."[15] Yet, despite the high number of prison religious conversions, a study has shown that belief in deities doesn't make a prisoner any less likely to commit new crimes after release.[16]

These are just a few of the many examples of how we can best reform our society such that religion isn't shoehorned into every facet of every program meant to improve lives or make people better.

Science Replacing Religion

For thousands of years, there have been people who feel that religions have been making promises they can't fulfill. From the claim that prayer can heal the sick to the promise we can live forever after death, religions offer the world to those filled with wishful thoughts, but leave some saying the traditions provide

---

Pew Research Center, March 22, 2012, https://www.pewresearch.org/religion/2012/03/22/prison-chaplains-exec/.

15. Mona Chalabi, "Are Prisoners Less Likely to Be Atheists?" *FiveThirtyEight*, March 12, 2015, https://fivethirtyeight.com/features/are-prisoners-less-likely-to-be-atheists/.

16. "Analysis of 174 inmates finds religion 'superfluous' to recovery when compared to factors such as access to safe housing and family support." Charles Hymas, "Prisoners Who Find God Just as Likely to Reoffend as Those Without Religion," *The Telegraph*, January 6, 2022, https://www.telegraph.co.uk/news/2022/01/06/prisoners-find-god-just-likely-reoffend-without-religion/.

little more than the occasional inspirational word. Scientific discoveries, however, have begun to fill that void and make good on many of religion's seemingly impossible offers. Whether you choose to believe that a deity inspired that science, or acknowledge the independent nature of human endeavors, it's important to understand what it can offer us.

An afterlife is something that most religions advertise, in one way or another. Because we have knowledge of our own impending death, humans are susceptible to this type of belief. But believing doesn't make it so … and to date, there's not a shred of evidence to support any idea of life after death. However, industrial designer Gerard Moline has come up with a way to combine that romantic notion with reality.

Moline's *Bios Urn*, made from biodegradable materials like coconut shell and compacted peat, allows you to live after death … as a tree. The urn contains a seed, and after you're cremated, your ashes join the seed underground. The ashes serve as nutrients to help the tree grow and to make sure you really can live on after death. You can even choose from a variety of seeds to determine what type of tree will sprout from your urn. There are countless other examples of ways to be forever connected with the earth, even after death.

Then, there are vaccines.

Christians believe, as is reported in the New Testament scriptures, that Jesus of Nazareth healed ten men with leprosy. It sounds like an astounding feat, and perhaps it is, but compare that to Jacinto Convit, who saved thousands of lives when he developed the vaccine that protects us from it. In 1988, Convit was nominated for a Nobel Prize in Medicine for his anti-leprosy vaccine. So, while the promise of Jesus' healing power is a centerpiece of the Christian myth, the demigod's results leave something to be desired when compared to the rigor of man's scientific inquiry.

It makes one wonder why Jesus himself wouldn't have provided the vaccine, or at least given hints in his teachings to cures for some of the most horrific diseases that have historically haunted mankind. Or maybe Jesus saw those 10 people as superior—more worthy of a healthy life—than anyone else who has suffered from leprosy before, during, and after that time.

Evolution is another example of this.

Creationism is the religious belief that life, the Earth, and the universe are all created by a supernatural being; Creationists argue that we are well-suited for Earth and that this points to the intervention of a higher power. But a scientific understanding of evolution by natural selection, pioneered by Charles Darwin, tells us exactly why we are well-suited for some aspects of the Earth … and not so well-suited for others.

Through scientific analysis, we can see how the gradual, non-random process of natural selection has worked, and how it allows biological traits to become more or less common in a population based on the reproduction of those who bear those traits. Religion promises perfection, and while humanity is nowhere near that benchmark, natural processes are working to shape humans and all living things to be the most successful versions of themselves, and science is helping us understand how.

In each case, religion has made a claim that natural processes and scientific inquiry have worked to fulfill. So, while religions make lofty promises about life after death, cures for deadly diseases, and beauty and perfection around us, bringing those ideals to fruition is our responsibility.

4 Science and Reason Save Us All Time and Money

"Compound interest is the eighth wonder of the world.
He who understands it, earns it …
he who doesn't … pays it."

—Albert Einstein

Science is really incredibly simple, and you don't have to be a researcher to take advantage of that process. Actually, the scientific method boils down to steps so easy a baby could carry them out: make an observation, form a hypothesis, perform an experiment, analyze the data, report your findings, and then invite others to reproduce the results.

The fact is that we all carry out this process nearly every day, whether or not we realize that's what's happening. A quality mechanic, for instance, likely takes a scientific approach to their work, disregarding automotive fixes that have been proven unreliable. Likewise, a successful salesperson will achieve more if they use psychology to their benefit. Consider this quote from author Jonathan C. Smith: *"When you test an idea, you often use sci-*

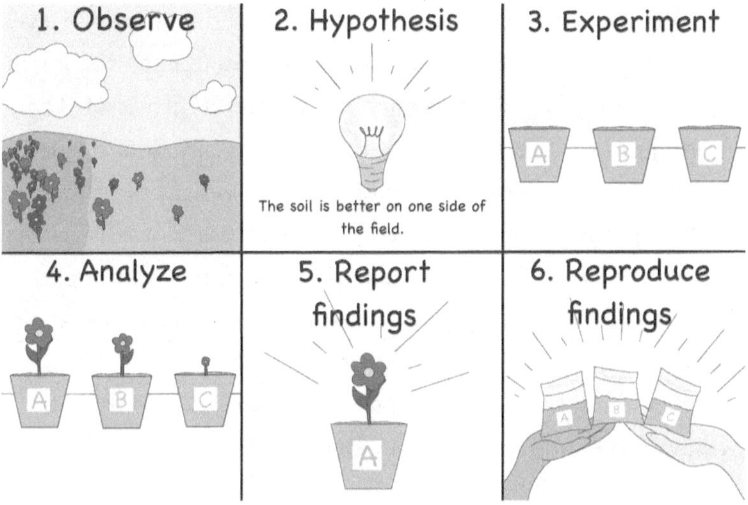

ence. When you try to find out what works and what doesn't work, you use science. When you take a pragmatic or practical approach to solving a problem, you use science. Doctors use science to diagnose illness and prescribe treatment. Detectives use science to solve crimes. Auto mechanics use science to fix cars. Students use science to decide on courses, career paths, and even weekend dates."[1]

Smith also added that sources and logic "can provide impressive support for a claim," but science is the ultimate arbiter of veracity.

"However, all are trumped by scientific observation, tests of hypotheses as well as explanations and theories," he added. "The famous expert can be proven wrong. A logical argument may be sound, but based on false premises."

1. Jonathan C. Smith, "Preface," *Pseudoscience and Extraordinary Claims of the Paranormal: A Critical Thinker's Toolkit* (Malden, MA: Wiley-Blackwell, 2010).

Science is so ever-present in our lives that we don't even realize when it's in use. Consider recreational activities, for instance. Saint Louis University mathematics professor Dylan Brody Johnson has compiled a list of all the ways that science—and specifically physics—is involved in bowling. On hitting the pocket, Johnson says, "The conventional wisdom tells us that strikes are more frequent when the ball enters the pocket at about a six-degree angle, measured relative to the centerline of the lane. A simple geometry problem confirms that this is impossible if the ball is rolled in a straight line, due to the narrow width of the lane. In fact, we can't do any better than two degrees with a straight-line path."

"The magic of bowling lies in the fact that a bowling ball with an appropriate amount and direction of spin can curve left or right and speed up or slow down as it slides down the lane," according to Johnson's analysis. "The trajectory of the ball in the lane is essentially parabolic and we can achieve much greater angles of impact on the pocket using parabolic paths. Six degrees is possible with a curved trajectory."[2]

It's clear science can help you do everything from fix a car to improve your bowling skills. To better understand this notion of using science to help you in daily life, here's a scenario:

> You are grocery shopping, and you see a new type of dessert that you want to try (Observation). Considering your past experience, you think that this treat will be perfect for your whole family (Hypothesis). You decide to buy it and see if

2. John Gonzales, "7 Ways That Physics and Science Relate to Bowling," *M Live*, January 20, 2017, https://www.mlive.com/dining/2017/01/7_ways_that_physics_and_other.html.

> you are right (Experiment). After dinner, you share the dessert with your family, and ask their thoughts (Analyze). You then tell your friends about the experience the next time they're shopping (Report and Reproduce).

Similarly, reason is something we use every day. Any time we use logic to make a decision, we are using *reason*. This is something we should *all* do more of. You might not realize it, but this is where philosophy meets the real world. Consider this situation:

> You are ordering pizza for yourself and two friends, a total of three people. One of those people—Eric, your roommate—is a vegetarian. You only have enough money to buy one pizza, but your friend Jenny from work loves barbecue chicken. You also prefer the chicken, so you have to decide whether you'll disappoint Eric or your colleague *and yourself.* Do you buy a cheese pizza, which will really only satisfy Eric, or do you go with the option that will help the greatest number of people? At this point, you have a moral dilemma not unlike the *Trolley Problem*. In that classic example, an individual is put in a situation in which they have to choose between letting a few people get hit by a trolley or pulling a lever that changes the track and hits just one person. In the situation above, using logic and reason, as well as your own emotions, can ultimately help you decide which pizza to order.

So, we've established that anyone can research, look at data, and draw conclusions. And we've shown that logic and reason can *actually better our lives*. So, why *wouldn't* we all want to use these processes to learn more and improve our lives?

If science and reason are universally established as the best methods for getting to the objective truth, which they are, then why should they be confined to college classrooms and high-end laboratories?[3] Think of all the ways science, reason, and critical thinking could improve your life: you could cut down on your bills by analyzing them and eliminating what you don't need, you could save time on your daily commute by trying different routes and then adjusting accordingly, and you could even accomplish more by critically evaluating your work patterns.

Sciencing Every Day

Let's take a look at some examples of how science can be used to improve daily life:

Driving a Car

Knowing elementary physics could help you to become a better driver, and therefore save money on maintenance, repairs, gas, tickets, and more, according to Dan Handley, Chief Scientific Officer at the Clinical and Translational Genome Research Institute.

---

3. "Scientific thinking is incredibly important, and the best part is you don't have to be a scientist to acknowledge that fact or to help others see it. Science is not some holy book that can only be interpreted by the Elder Scientists. It is not some club that denies entry to outsiders. It is a system of discovery that we've all been practicing informally since we were children. It started the first time we noticed something unusual, asked a question about it, and then put it to test." David G. McAfee, *No Sacred Cows: Investigating Myths, Cults, and the Supernatural.*

When he is driving, he thinks it is important to understand "*potential and kinetic energy going up and down a hill, the concepts of inertia, friction, and acceleration, and* ***their consequences on fuel economy***," Handley told me in an interview.

Science can also come into play when choosing a route, as discussed above. If you want to see which trip is faster, smoother, or less plagued by traffic (your hypothesis), you have to test it (that's your experiment). Once you conduct this test several times under varying conditions, accounting for those changes as best you can, you might find that one path is easier on weekends but that weekdays are different. Now, on any given day, you can choose which way to travel based on your particular needs at that time. And you can even share this information with your friends and family, allowing them to duplicate your scientific findings. If this sounds familiar, it could be because *you already do this*. I think most of us do. The trick is to do it more often, in more situations, and with more awareness of what you're actually accomplishing.

Paying Your Bills

When I was in the first grade and living with my grandparents, my grandpa had a controversial hypothesis. He thought that, if he could encourage me to start turning off lights whenever they weren't in use, he could save money on his monthly electricity bill. To test this notion, he offered me a small monetary incentive, proportionally based on the savings. In other words, he offered to pay me 50% of the savings on our average monthly bill … nothing more than cheese to get the mouse through the maze. He even gave me the title of "Light Monitor" to make it official, which, in hindsight, was probably his attempt to make me take it even more seriously. And it worked.

After the deal, my grandparents' monthly electricity bill went down significantly—from an average of $40 to $22—and I pocketed a little cash. My grandpa nearly cut his bill in half, saving about $216 per year, and his experiment was successful. Later, I heard him telling a friend about the deal, effectively completing the scientific process.

My grandpa certainly used the scientific method to save money on his bills, but his isn't the only way to do just that. If you have a different hypothesis, like that your family would be happy without cable if you had more streaming services, then you should experiment with it. Critical analysis is important for cutting unnecessary expenses and for budgeting in general.

Corporal Punishment

Fortunately, taking advantage of the scientific method doesn't always involve *direct* experimentation. You can also indirectly benefit from the process by valuing the scientific consensus. Take corporal punishment, for example. Most people probably don't approach the subject of child discipline from a scientific standpoint, but that doesn't mean it can't be done. And it should certainly be done more often, despite the inherently emotional nature of the topic.

I would never *tell* anyone how to raise their children, except when it violates laws and ethics, but there *is* a scientific approach to the controversial issue of physical discipline. All you have to do is pay attention to peer-reviewed scientific studies, which consistently find that deliberately inflicting pain on children as retribution doesn't usually work as a long-term solution to poor behavior. According to one in-depth analysis of 88 high-quality scientific studies on the subject of corporal punishment, unnecessary physical punishment can have negative effects on moral internalization, quality of parental relationships, mental health, and even social behaviors.[4]

That scientific consensus, which represents unity behind solid, repeatable data, is further explained in a 2012 feature article for the American Psychological Association:

"It's a very controversial area even though the **research is extremely telling and very clear and consistent about the negative effects on children**," says Sandra Graham-Bermann, PhD, a psychology professor and principal investigator for the Child Violence and Trauma Laboratory at the University of Michigan. "People get frustrated and hit their kids. Maybe they don't see there are other options."

Many studies have shown that physical punishment—including spanking, hitting and other means of causing pain—can lead to **increased aggression, antisocial behavior, physical injury and mental health problems for children**. Americans' acceptance of physical punishment has declined since the 1960s, yet surveys show that two-thirds of Americans still approve of parents spanking their kids.

But spanking doesn't work, says Alan Kazdin, PhD, a Yale University psychology professor and director of the Yale Parenting Center and Child Conduct Clinic. "You **cannot punish out these behaviors that you do not want**," says Kazdin,

4. Elizabeth Thompson Gershoff, "Corporal Punishment by Parents and Associated Child Behaviors and Experiences: A Meta-Analytic and Theoretical Review," *Psychological Bulletin* 128, no. 4 (2002): 539.

who served as APA president in 2008. "There is no need for corporal punishment based on the research. We are not giving up an effective technique. We are saying **this is a horrible thing that does not work**."[5]

Studying the robust evidence on this matter and incorporating the lessons into your own life could save you time because you'd be helping your kids learn lasting lessons instead of merely instilling fear in them. By seeking alternative punishments, you could prevent having to do it all over again next time, and maybe even foster a better relationship with your child. That's not to mention all the money you could save on doctor bills and help with their mental health.

If you don't have time to perform tests and you don't have access to peer-reviewed studies, but you still want to use scientific principles in your daily life, that works, too! Here's what I mean:

Buying a Product

Every single time we buy anything, we have an opportunity to use concepts that are key to science. Depending on the level of your investment, you may decide to do your due diligence in any number of ways, but whether you're shopping around for a new computer or a mortgage, it's important to analyze all available data. Let's say you need to buy a new TV because yours is broken. Do you just walk into the store and buy the first one you see? Not if you value critical thinking, a key component of the scientific method. If you want to be scientific about it, you'll

---

5. Brendan L. Smith, "The Case Against Spanking," *American Psychological Association* 43, no. 4 (2012): 60.

look at reviews, check stats, measure the space, and make an educated decision. This will likely save you money and make sure you're happier with your purchase in the long run.

Scientific principles can even come into play when you're buying medicine from a pharmacy, according to Dr. Anastasia Bodnar, a science communicator with expertise in plant genetics and sustainable agriculture.

"When it comes to medicine, there are a lot of competing products for a given symptom. If you have a cough, you can get lozenges, sprays, capsules, liquids. **Some of the active ingredients have been shown to work in clinical trials for certain symptoms, and some have not**," she told me in an interview. Dr. Bodnar added that some claims are "*completely bogus*." "How can we be sure to get the product that is going to work, and not waste our money? It's a combination of science and trust," she noted. "When possible, **I do look up studies on different ingredients and find what actually works. I avoid products with impossible claims, such as homeopathic products**."

You can also perform tests on a more personal level, Dr. Bodnar said. "For some products like lozenges and sprays, it's just a matter of **trying them and remembering what works best for me**—for example, our family prefers lozenges with pectin to coat the throat. But I can't always find information on my own, so that's where trust comes in. I **seek medical professionals, including pharmacists, who have an attitude of, '*I'm not sure, let's look that up.*'** I want them to **be inquisitive and not pretend they know everything**, especially with newer products. Then I know I can reasonably trust their evaluation of a product."

Kitchen Science

A lot of people will tell you they don't like or understand science, yet will bake up a perfect pie every time. What they might not realize is that science is key to baking and cooking in general. If you excel at baking, for instance, you might do well in chemistry classes. You aren't just mixing baking soda and yeast and sugar; you are, in fact, creating different chemical reactions in order to make the product rise perfectly. And in either case, you need to be familiar with the Maillard reaction, which *Merriam-Webster* defines as "a nonenzymatic reaction between sugars and proteins that occurs upon heating and that produces browning of some foods (such as meat and bread)." In other words, it's the heated browning of your dish.

But that's not all. Consider the simple act of cooking rice at high altitudes. If you don't know anything about science, you're not going to be very successful at it. I moved to a beautiful mountain town years ago, and couldn't figure out why I couldn't seem to cook rice. I had done it successfully when living at sea level, but had no idea my new lake house would affect that process. I soon learned that you have to add an extra minute of cooking time for every 1,000 feet above sea level, because the boiling point of water is lower up high, and it therefore makes things take longer to cook.

Cleaning Up

It may sound crazy, but scientific principles can even be used to save time and money when it comes to projects as simple as cleaning up the house. Consider loading up the dishwasher. If you do this task every day for several years, you may start to notice that certain positions and techniques work differently with your various dishes on that particular appliance. If you have a scientific mind, you might start to experiment with how your dishes fit and how they're placed until you find an arrangement that works best. By using the most effective placement for every single item, you cut down on your use of dish soap and even water consumption.

The American Cleaning Institute, which says it has a "continuous commitment to sound science and being a credible voice for the $57.4 billion cleaning products industry,"[6] suggests you "load dishes along the back and sides so that they face the middle of the dishwasher."

"Glassware, coffee mugs and plastic ware should go on the top rack. There is higher water pressure on the bottom which could break some glasses, and the water on the bottom is higher temperatures, which could melt the plastic," the Institute says in providing another potential money-saving tip.[7]

To save time, you can also apply these principles to *how* you clean. Do you just start picking things up, or do you make a plan to cover the most ground in the least time? Do you consider planning so that you don't have to walk over areas where you've already mopped, thereby keeping you from having to do it all over again?

---

6. "ACI Jobs," American Cleaning Institute, cleaninginstitute.org/about-aci/aci-jobs.
7. "Loading the Dishwasher," American Cleaning Institute, https://www.cleaninginstitute.org/cleaning-tips/dishes/loading-dishwasher.

Principles in Action

We've gone over several specific daily tasks and how we can use science and reason to improve how we do them, but it may also be helpful to work backward to get an even clearer view. If we look at common scientific and philosophical principles, which are usually only taught in an academic setting, we may be able to see how we can apply those to life in order to change things for the better.

The Socratic Method

The Socratic Method isn't a direct part of the scientific method, yet they are related. And they're both helpful in various circumstances. The easiest way to think about it is that the scientific method is focused on *testing*, while the Socratic Method emphasizes *questioning*. And questioning just happens to be a key component of all scientific endeavors.

The Socratic Method is basically a form of advanced debate that isn't necessarily meant to prove that someone is wrong. Instead, the intention is to ask questions that will uncover the truth and perhaps make your interlocutor discover flaws in their own argument. This is something that can definitely come in handy in daily life.

Imagine being able to diffuse arguments and win debates without ever looking like you are trying to debate in the first place. Picture yourself coming out on top in verbal interactions, without ever having to attack anyone, or even form an opinion yourself. All this could be yours if you follow the Socratic Method, which is basically just asking questions and follow-ups until you reveal a potential contradiction in someone's thinking, and then working together to resolve it.

Here is an example I created with the help of philosopher Peter Boghossian to show you how to use the Socratic Method in real conversations. In this case, I used a variation of the Socratic Method and applied it to a high-pressure sales situation regarding a multi-level marketing (MLM) scheme.

Person 1: Hey, buddy! Do you ever think you need some more spending money, fast? I have an opportunity for you.[8]

Person 2: I'm actually quite content. Thanks![9]

Person 1: This is completely different than anything you've ever heard of, and I promise it'll be worth your while. Have I ever steered you wrong?[10]

Person 2: Well, I'm skeptical, but I'll hear you out. I just need to know if it really works. I'm sure you can understand that. How much have you made? How long have you been doing it?[11]

Person 1: Sure, I understand. I haven't started making the big bucks yet, but I will once I pay my certification fees. I've

8. The conversation begins in the "wonder" stage of the Socratic Method dialogue structure, with person 1 as the initial inquirer. That's because these types of MLM schemes also utilize questioning in their scripts.
9. In this part of the dialogue, person 2 offers their hypothesis: the opportunity isn't necessary.
10. In this part of the example, person 1 is pushing back against person 2's hypothesis by providing counterexamples.
11. Because this is based on a familiar interaction, person 2 revise's their previous position, and decides to hear out person 1's pitch.

only been doing this for six months. It's a perfect system that always works because we provide products and services that people actually want!

Person 2: You said this would be a "fast" way to make money, and you haven't done so in six months. That doesn't make sense to me. What do you mean by "fast"?

Person 1: One year.

Person 2: Can you tell me how the company works?[12]

Person 1: You make money based on the sales of the people you sign up! It's a never-ending stream once you get going.[13]

Person 2: It sounds to me like the sign-ups are more important than the products and services, and you make the most money from getting others to become salespeople themselves, which is similar to a multi-level marketing company. Does all that sound about right?[14]

Person 1: Just about, but there's more to it than that.

Person 2: So, you agree that the structure of this company is basically similar to that of a multi-level marketing organization, which is almost always a get-rich-quick scheme that

---

12. Because of the cyclical nature of these conversations, at this point the discussion restarts, with person 2 wondering how the company works.
13. Here, person 1 offers a response in their own words.
14. This section is inspired by the first of Rapoport's Rules for ensuring productive dialogue. The rule is often quoted in the formation provided by Daniel Dennett: "You should attempt to re-express your target's position so clearly, vividly, and fairly that your target says, 'Thanks, I wish I'd thought of putting it that way.'"

doesn't pan out for most people. Do you agree there?

Person 1: This isn't a get-rich-quick scheme at all. It's a legitimate way to make money and help others make money at the same time!

Person 2: Well, since you make money from signing other individuals up as salespeople, and you haven't made much in your six months working there, I think my skepticism is warranted. If I told you to give me $5,000 today, and that I'd double it for you in a month, would you do it without evidence?[15]

Person 1: Well, no ...[16]

Person 2: Then I think you can understand. I trust you, but my investment of time or money is important to me, and since you've not made anything in six months, I'm asking of you what you'd ask of me if I told you to give me $5000 and let me double that for you.

Person 1: Hmm. Yeah, good point. You've given me a lot to think about![17]

In this example, which represents a modified form of the Socratic Method that's better suited for casual conversations, we were able to show how you can save yourself time and money through nothing more than questioning. There are actually several mechanisms for using philosophical concepts, critical think-

15. After a few lines of questioning and establishing definitions to various terms, person 2 offers a concrete counter-example.
16. Here, we elicited a contradiction.
17. It doesn't *always* happen, but often this type of discussion can lead to a revision of beliefs.

ing, and good old-fashioned analysis to prevent you from being tricked and keep money in your pocket, and we'll address them in detail.

Occam's Razor

Occam's Razor is an extremely simple yet effective philosophical concept aimed at uncovering the truth with very little information. Named after the famed philosopher William of Ockham, the principle states that the hypothesis with the fewest assumptions is likely to be correct. This is traditionally used in the context of philosophy, but it can be applied to just about anything. And like the Socratic Method, it can be used to keep you from being tricked and therefore save you money.

> **Internet Scams**. Let's say you get an email from someone who claims to be a prince in Africa, and who says they want to send you an obscene amount of money in exchange for a minor task. They say they'll send you a check for $100,000, and all you have to do is cash it and send half back. Without looking further into this, a plan like this appeals to many people. Assuming the "prince" is telling you the truth, you get a major reward for very little work. But what happens when we analyze this with Occam's Razor? If you try to justify the request as legitimate, you'll run into all sorts of obstacles and assumptions. First, you have to explain why a prince would give you $50,000 for basically nothing, and then why he couldn't just ask one of his trusted servants to do it. Why couldn't this "prince" just go to a bank, where the management would do basically anything for his busi-

ness? Finally, you'd have to answer why this royal representative would have contacted *you* above other people. That's a difficult one.

But there's a much simpler explanation than the one presented above. The answer that makes the fewest assumptions is that this person is not a prince at all and that they are trying to scam you. They likely send that email to thousands of different people, hoping a few of them will take the bait and deliver them a payday.

Psychic Advice. Let's look at a different situation. In this case, you get a phone call from an unknown person who says they can tell that you're going through stressful times, and that they can help. They say they felt spiritually compelled to contact you. Perhaps you *have* been experiencing something trying, so it might seem like a good idea at first. Then they say it costs $2.99 per minute. If you believe this person is legit, then you have to make several assumptions, including that "psychics" exist despite a consensus of scientific evidence indicating they do not. Then, you have to believe this person intuited your phone number out of nowhere, on the basis that you were having "stress," which most people experience often. You also have to believe that, for some reason, you were chosen over a nearly unlimited number of other people.

The more likely explanation, because it requires way fewer assumptions, is that this "psychic" is using a phone book to look for clients.

Peer Review

In science, peer review is a structured process that allows competing scientists to disprove (or confirm) the work of others through rigorous tests. Often, this process leads to successful repetitions of an experiment, further solidifying the hypothesis, but the whole point is to look for flaws and debunk the idea. The system was created to encourage or incentivize scientists to disprove others in their field. This can *also* be applied to life.

The next time you make a major decision, consider replicating this review process by asking your friends or family to look at your take with a critical eye. I actually do this all the time in my writing. I'll tell my wife an idea, or send her a draft, and ask her to look for what other people might have a problem with. She specifically looks for ways to poke holes in my theories, and to misinterpret them, so that I can improve my work. She has even reviewed this sentence that you're reading now.

The principles behind the peer review system can also be used when it comes to making decisions on what electronics to buy, like new headphones, according to Michael Simpson, who has 30 years of experience in endocrinology and cardiology. "I kind of engage in a form of systematic review of comments and reviews of a product. I will **give more weight to lengthy in-depth reviews, and less weight to short reviews, either positive or negative, that don't give me much information**," Simpson told me. "I use Yelp all the time in that manner to choose a restaurant or dentist. I make my judgment on this systematic review by giving more weight or attention **to better reviews and higher quantity reviews**."

I spoke with retired biologist Gina Melton about these issues and more. Melton, who said she uses "science in indirect ways" now that she's retired, confirmed it's a money saver for her.

"Someone said that science is a good BS detector, and you tend to use logic and critical thinking skills readily," she told me in an interview. "It helps to budget money and think out of the box. Also, I'm a very good cook because recipes are a lot like experiments."

Bird's-Eye View

Let's look at this on a larger scale. Keeping a few bucks in your pocket is one thing, but saving millions from all Americans is quite another. I'm talking about the fact that, if we emphasize science and reason-focused thinking on the biggest scale possible, we can reform on an unprecedented level. I previously founded a political non-profit called the Party of Reason and Progress, which created a platform to enact an executive agenda completely based on valid data, using only peer-reviewed, published studies to support policies. Part of that platform involved the cutting of programs that haven't been proven to work or perhaps have even been proven *not* to work. There are countless examples of these types of spending, but let's look at what might be the most outrageous one.

The U.S. government has spent $20 million to pay so-called psychics to find things for them, such as bombs, targets, and water. Some of this included literally paying people who claimed to be psychics as much as $500,000 per year, and it also included paying for classes to try to teach troops to do things like "remote viewing," which is a debunked method for finding something by imagining geological features or even painting the area where the item or target might be.

"Sketches were not always on target. To no avail, one set of Pentagon planners consulted psychics to pinpoint where Colonel Muammar Gaddafi was staying before U.S. warplanes attacked Libya in 1986. Another intelligence unit asked psychics to picture where an agent suspected of being a double stashed the money he made spying for the other side. (They could not say)," according to a Forbes report from 1995, when the program was shuttered.[18]

According to the report, "Actually, the study came up with an estimate: Star Gate psychics got it right only about 25% of the time." It quoted a CIA study that found that the project produced "a large amount of irrelevant, often erroneous information."

The article also quotes Jessica Utts, a statistician at the University of California at Davis who was involved with the CIA study, as saying, "Sometimes it seems that these people are right on … But nobody knows when those times come."

These are just a few examples of how science and philosophy can be used to better your life (and in some cases, *our* lives), and more can be discovered quite easily. In the next chapter, we look at the beauty that can come along with living an evidence-based life.

---

18. Douglas Waller, "The Vision Thing," *Time*, December 11, 1995, https://web.archive.org/web/20070209085903/http:/www.time.com/time/magazine/article/0,9171,983829,00.html.

5 More Beautiful Than Fiction

"Reality leaves a lot to the imagination."

—John Lennon

They say that reality is sometimes stranger than fiction. I completely agree, and I think the real world is often *more beautiful* than fiction, as well.

Don't get me wrong. There is no shortage of ugliness on planet Earth; that's for sure. But the beauty that's inherent within it is often underrated or unappreciated. Sometimes you have to look a little closer to see it, but that doesn't mean it's not there. And while some people believe science and reason detract from that aesthetic, others theorize that they can actually help one *see* it. This discovery can even be an emotional or *spiritual* experience.

We talked about how famed astronomer Carl Sagan, a hero to scientists and the world, once said,

> Science is not only compatible with spirituality; it is a profound source of spirituality. When we recognize our place in an immensity of light years and in the passage of ages, when we grasp the intricacy, beauty and subtlety of life, then **that**

> **soaring feeling, that sense of elation and humility combined, is surely spiritual.**

For Sagan, an understanding of science helped intensify his appreciation for the natural world. In other words, the world was made *more beautiful* through the lens of science. But Sagan's spiritual views don't represent the *only* way to experience this unique feeling.

Life is actually "*more colorful and fun with a dash of science*," according to Dr. Anna Zakrisson, a science communicator who has five degrees, including a PhD in microbial ecology. "The natural world is amazing beyond belief. **I sometimes find myself staring in sheer wonder on my own hand, or a leaf on a tree**," Dr. Zakrisson told me. "I feel **that the scientific method is a way of us humans to objectively connect with each other over matters related to this physical world**. We can define that the leaf is green by determining the frequencies of light that are bounced off from its surface."

Like Sagan, Dr. Zakrisson sees science as an emotionally charged subject. She also believes art and science can work together under certain circumstances.

During our conversation, Dr. Zakrisson said, "I'm fully convinced that we **need the arts to communicate science**; to emotionally catch people; to motivate them to learn the language of science. I disagree that science is emotionless, gray, and cold. To me, **science is extremely emotional**; it is a never-ending childish curiosity. Science is **uncomfortable and revolutionary**. It is euphoric and horrific because it also means accepting, and publicly acknowledging, those **times when I might have talked out of my arse**."

This perspective is shared by neurologist and psychiatrist Jonathan Howard, who said in an interview with me that his medical passion stems from a fascination with a "three-pound mass in our head" that "contains our entire universe." He argues that, on one level, the brain is pretty dull. But that is only part of the story.

"Seeing a brain is an amazing experience, in that someone's entire memories, experiences, and sensations are in your grasp. **But few people would say this lump of dull, yellow tissue is *beautiful*,**" Howard said. "Yet, **the brain can be an object of incredible beauty**."

Howard's office is decorated with several types of pictures of the brain, including historical anatomical drawings that he's inspired by. "I **find it inspiring that my predecessors took so much time and effort to artfully depict an organ whose workings they knew so little about**. It was clear these early anatomists also took care to make their work beautiful."

Howard also has art featuring neuroanatomical images from MRIs and other brain scans, as well as microscopic images of brains.

Art Merges with Science

Art and beauty actually intersect with science in several ways, including in the work of legendary scientist and talented artist Maria Sibylla Merian. If you haven't ever heard of her, that's completely understandable. She isn't a household name, but her story is epic.

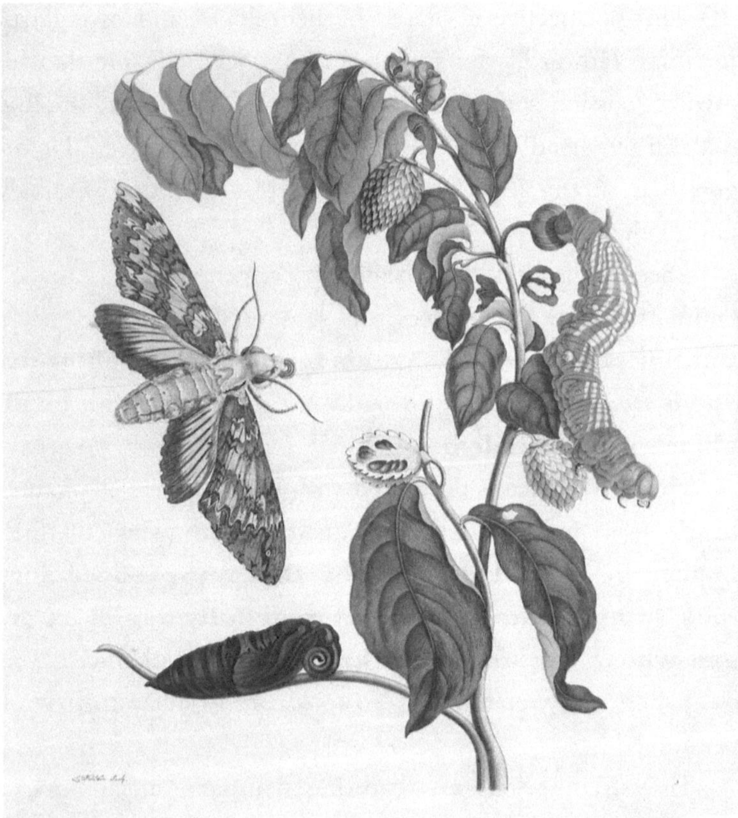

Credit: Maria Sibylla Merian; engraved by I. Mulder - Maria Sibylla Merian, 1705. *Veranderingen der Surinaamsche Insecten* via Wikimedia Commons.

I refer to Merian as a scientist because that's how I see her, and how I believe she likely saw herself. That being said, she was born in 1647, at a time when women were actually considered inferior and *incapable of scientific pursuits*. It was believed that the rigorous life of a researcher would interfere with the "domestic duties" of women.[1] As a result of the restrictive time period in which she lived, Merian decided to become an artist, but there's no doubt that she had the mind of a scientist.

At just 13 years old, Merian started drawing plants and insects, including silkworms. She became the first person to record the lifecycles of butterflies and moths—and probably one of the first ecologists—by using her art to record scientific observations. She didn't have any formal scientific training, but she did have curiosity. And her work still influences scientists today.

Merian's story should be an inspiration to anyone who isn't a formal scientist yet has that innate curiosity that drives all discovery. She shows that we can all think like scientists—because we can all research, look at data, and draw conclusions—and that doing so doesn't have to detract from the beauty in the world. What I'm saying is it's entirely possible to have your cake and know the science that went into baking it, too.

For Merian, her passion for the natural world manifested in her study of silkworms and caterpillars, and she produced amazing and informative art as a result. She was fascinated by these little, creepy crawlies that others found disgusting, and her scientific curiosity merged with her art to create something even more beautiful. You may not think silkworms are as incredible as Merian did, but knowledge of how things work can make *anything* more beautiful, if it's properly applied.

Merian isn't the only one who has merged art and science. About 200 years after Merian's work, but still before the advent of macro photography, Ernst Heinrich Haeckel stepped onto

1. Thankfully, in this day and age, absolutely nobody still thinks this way. /Sarcasm

the scene. A German zoologist and proponent of Darwinism, Haeckel drew life as he saw it through his microscope.[2] The results were incredible. His drawings and watercolors were never his career, although he did consider them as a profession in his younger years. Instead, his art helped him communicate to the world more about evolution by natural selection and its many possible implications.

Haeckel was pressured into a medical degree by his family, and after practicing medicine for a short period of time, he traveled to Italy, where he painted. It is after this artistic venture that Haeckel started practicing zoology. But that artistic spirit never left him, as he ended up producing some of the most beautiful representations of nature ever recorded. He went on to influence the Art Nouveau artistic movement, which is known for featuring flowers, leaves, animals, and insects, according to an educational website promoting his work.[3]

Haeckel is said to have "*devoted his life to the study of flora and fauna from the highest mountains to the deepest oceans*."

"He discovered, described, and named thousands of new species, mapped a genealogical tree relating all life forms, and introduced many terms in biology, including ecology, phylum, phylogeny and Protista," according to the educational site's biography. "The published artwork of Haeckel includes over 100 detailed, multi-colour illustrations of animals and sea creatures."

2. Gloria Robinson, "Ernst Haeckel, German Embryologist," *Britannica*, February 12, 2026, https://www.britannica.com/biography/Ernst-Haeckel.

3. "Ernst Heinrich Philipp August Haeckel (1834–1919)," https://haeckel.tilda.ws/en#haeckel.

Understanding Increases Appreciation

In my opinion, and in that of countless scientists, academics, science writers, and communicators, understanding the nature of something beautiful doesn't take away from the beauty. Knowing about how a rainbow is formed doesn't keep one from stopping in their tracks when they see the magnificent colors stretching over the sky in the distance. And in many ways, comprehending the intricacies of light refraction can help us appreciate the experience even more. It's no longer a mysterious bundle of pretty colors; now the rainbow is a diverse spectrum of light being reflected in every direction. The mystery may be gone temporarily, but in its place, you are left with the never-ending rabbit hole of inquiry. And when we are graced with a *double* rainbow, my awe is compounded by the knowledge that it's created when "sunlight is reflected twice within a raindrop with the violet light that reaches the observer's eye coming from the higher raindrops and the red light from lower raindrops," according to the United Kingdom's national weather and climate service.[4]

Not only does scientific knowledge help some things feel even more amazing, but sometimes an understanding of something is all it takes to pique an interest and change someone's entire perspective.

When I was learning to drive, my mom wanted to teach me to drive a stick before trying anything else. She felt it was important that I learn to drive a manual transmission, just in case there was an emergency and I had to do just that. She took me to a

---

4. "What Is a Double Rainbow?," Met Office, https://weather.metoffice.gov.uk/learn-about/weather/optical-effects/rainbows/double-rainbows.

commercial area with little traffic and put me in the driver's seat. Her first three children quickly and easily learned to drive that way, but I didn't want to drive a stick, and I didn't understand how it worked. After I stalled the car a few times and threatened to quit, my mom had a brilliant idea.

She knew I wasn't like her other kids, in that I always had to understand how things worked. She decided to walk me through *how* a manual transmission works, explaining that transmissions exist because of the physics of gasoline engines. She told me that you push down the clutch so that the engine can be separated from the drive wheels while shifting. She even drew a diagram that helped me comprehend the components and what I was supposed to do. After that, I drove us home with no problem, and never stalled another car again.

In other words, an understanding of the mechanical nature allowed me to excel at the task and even appreciate the overall experience more. Now, I love driving stick, and I understand the *reason* behind it. Years later, I uncovered another mystery: the reason I was able to drive a car with a manual transmission after receiving a thorough explanation of how they functioned. As it turns out, I have an analytical, logical learning style. According to the Critical Thinking Child LLC, which provides "academic tutoring, test prep, parent coaching, and teacher training to foster students' academic achievement," the analytical learner is also nicknamed the "scientist" or the "investigator." The site notes that "analytical learners love logic, puzzles and uncovering how their world works," and adds, "These learners sometimes drive their parents and teachers crazy by asking 'why?' all the time! They love being presented with new problems and working through them in a systematic way."[5]

But regardless of your learning style, understanding the science behind how something works will likely help you improve on how to put it in action. And you can take this lesson into your everyday life on just about any possible level. The next time you have trouble accomplishing something or learning a new task, feel free to try this technique. All you have to do is learn everything you can about the topic, including how and why it works and any scientific principles in action, and use that information to improve your abilities.

My personal story is only representative of my own experience, but it is further supported by Aryel Ripp, a bioanalytical chemist in the pharmaceutical industry. She says science doesn't take away from the beauty in the world; it actually adds to it.

"To me, **understanding why something is the way it is (scientifically) adds a different aspect of beauty**. For example, why are plants green? Why is the sky blue?" Ripp said in an interview with me. "I have always asked those basic questions even as a child."

Beauty in Nature

There is no doubt that nature itself produces some of the most breathtaking images. From the awesome Grand Canyon to a simple waterfall, it's easy to find beauty on Earth. But how much of that natural beauty you can see depends on how hard you're

5. Lemi-Ola Erinkitola, "Learning Style Analysis: Analytical Learners," The Critical Thinking Child, July 7, 2018, https://www.thecriticalthinkingchild.com/learning-style-analysis-analytical-learners/.

willing to work to understand.

First, let's address this literally. It was through science and rational thought that glasses, telescopes, and microscopes were invented, which then allowed us to see more of just about everything we call beautiful. Using the naked eye, there's only a certain number of things we're able to see and recognize. By opening up entirely new avenues of vision, we can now appreciate natural beauty on any scale, from microscopic to planetary wonders.

Without science or reason, for instance, stars would just be white dots in the night sky. They may even appear as a simple blanket of nightlights for our comfort. With our current technology and understanding, however, we can see an entirely different world filled with gases and fire. We learn that the stars are not perfectly aligned; they are actually far apart and often hold little to no relation to one another. We see that the universe is much more complex, and therefore more intriguing, than it might seem at first glance. Consider the below image of the spi-

Credit: NASA/JPL-Caltech/ESA/Harvard-Smithsonian CfA.

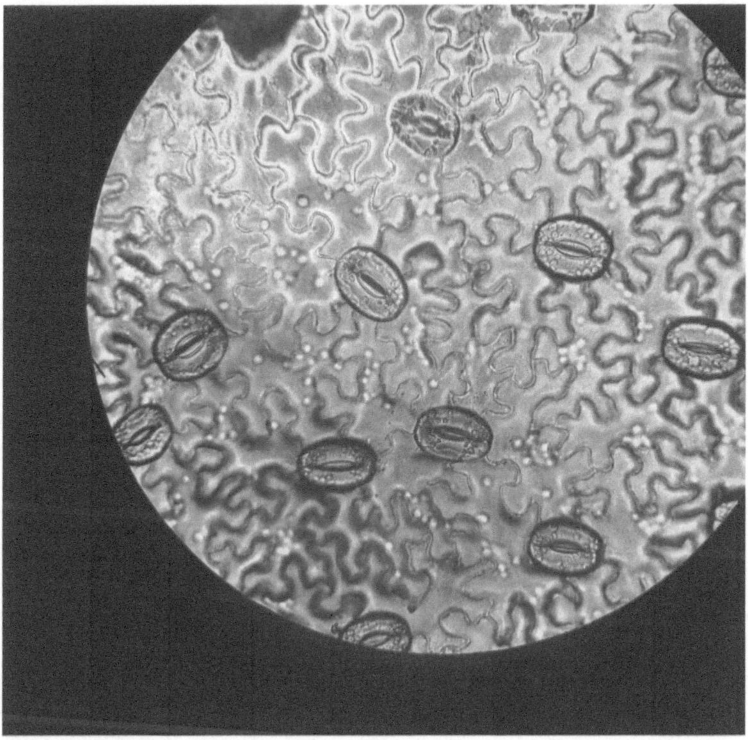

Credit: Tyanna, "Cuticle of Leaf under Microscope," November 6, 2015.

ral galaxy Messier 81, one of the brightest galaxies that can be seen from Earth through telescopes, according to NASA.

This same educational principle goes for both Goliath and David. On the opposite side of the spectrum, we have the world of microbes. We have the technology to see microscopic items and beings that otherwise would have remained mysterious forever. Above is a photo of a cuticle of a leaf under a microscope, provided by a contributor named Tyanna.

Another example is published in *Chemical & Engineering News*

Magazine.[6] In a 2015 article, the magazine featured an incredible image of "happy little plant cells." The cells are actually from a piece of grass, and it looks as if there are dozens of smiling faces.

"Those grinning faces are actually veinlike structures in a piece of grass. Phil Gates of Durham University created the glowing smiles in this micrograph by adding two dyes to a thin slice of a marram grass (*Ammophila arenaria*) and then shining blue-violet light on it. The first dye, calcofluor M2R, fluoresces blue when bound to cellulose," Manny Morone wrote. "The second, Auramine O, fluoresces yellow when it binds to the plant polymers cutin and lignin. The blue smiles are made of phloem tissue, which carries nutrients around the plant and has pure cellulose cell walls. Chlorophyll-rich cells, which fluoresce brownish-red without any added dyes, can be exposed as the grooves in this sample open up in response to water in the air, uncurling the blades of grass and triggering photosynthesis."

The important point here is that the scale is irrelevant. Whether you're looking at the Milky Way galaxy through a telescope or a microbe through a microscope, the beauty is there in reality. You just have to search for it. And science is a great way to do exactly that.

Perhaps most importantly, science and critical thought help us better understand the world, which always comes with its own satisfaction. By expanding our understanding of everything, from simple microorganisms to the universe itself, we give ourselves a deeper appreciation for what we have in life. Knowing

---

6. The image of these smiling faces can be found in the article "Chemistry In Pictures," *Chemical & Engineering News*, November 16, 2015, https://cen.acs.org/articles/93/i45/Chemistry-Pictures.html.

the chaotic nature of the universe, and that the Earth is just a ball of rock and water flying through space at incredible speeds, can even help us focus more on those things we may be taking for granted. Understanding that a rogue piece of space debris could put an end to this experiment called Earth at any time should encourage you to spend your time being happy with those you love. It might even make you realize how unimportant some of your daily gripes are. After all, how much does it matter that Janet ate your yogurt out of the fridge in the breakroom, in the grand scheme of things?

A Closer Look

The fact is that, if you look closely at anything in nature, science and logic can jump out at you.

Take a beautiful flower, for instance. You probably admire it for how it looks, and for how it smells, right? But how did our eyes evolve to see that color? And what is the chemical composition that causes the delightful perfume? You might even ask yourself *why* flowers are so often brightly colored, and why they smell so good in the first place. If you think about it logically for a minute, you'll probably come to a good answer: they have evolved to attract certain pollinators. But what about these thorns on the side of this rose that just pricked your thumb? "Every rose has its thorn," right? Wrong. Plant scientists distinguish between thorns—created from modified branches—and spines—created from modified leaves—and rose bushes have neither! They have "prickles." But "Every rose has its prickles" isn't as catchy. Still, I think a rose's prickles are more beautiful knowing all of this, as well as understanding that it evolved to keep herbivores from eating the plant. It's all about survival of the fittest, which was theorized by Charles Darwin. And being aware of that doesn't take away from the rose's beauty; it simply gives us another dimension in which to appreciate it.

This sentiment was also echoed by Ripp: "**Understanding the basis of why something is beautiful is exciting and amazing to me**. Why look at something beautiful and not ask why? Are people really satisfied with just the beauty?"

This is something that theoretical physicist Richard P. Feynman has addressed, as well. He not only talked about how a scientific view of certain things doesn't detract from their inherent beauty, but also used the same example of the flower:

> I have a friend who's an artist and has sometimes taken a view which I don't agree with very well. He'll **hold up a flower and say, 'Look how beautiful it is,' and I'll agree**. Then he says, **'I as an artist can see how beautiful this is but you as a scientist take this all apart and it becomes a dull thing,'** and I think that he's kind of nutty. First of all, the beauty that he sees is **available to other people and to me, too,** I believe …
>
> I can appreciate the beauty of a flower. At the same time, **I see much more about the flower than he sees.** I could imagine the cells in there, the complicated actions inside, which also have a beauty. I mean it's not just beauty at this dimension, at one centimeter; there's also beauty at smaller dimensions, the inner structure, also the processes. The fact that the colors in the flower evolved in order to attract insects to pollinate it is interesting; it means that insects can see the color. It adds a question: **does this aesthetic sense also exist in the lower forms?** Why is it aesthetic? All kinds of

A dragon rendered in neon lights displayed on the exterior of the Museum of Neon Art in Glendale, CA.

Credit: Andrew Keenan Richardson, March 17, 2016

> interesting questions which the science knowledge only adds to the excitement, the **mystery and the awe of a flower.** It only adds. I **don't understand how it subtracts.**

Now for a not-so-common example. Do you remember the last time you saw a neon sign? You were probably walking down a street, and you spotted one in the window of a bar or restaurant. Do you know how it was created? Those bright colors didn't come out of nowhere; it's chemistry. In fact, neon signs work by using electrical currents to excite a gas at low partial pressure. This is called *line spectra*, and it has allowed for the creation of some incredible art. Above is a dragon rendered in neon lights on display outside the Museum of Neon Art in Glendale, California.

Science and reason can also help you understand your fears,

your weaknesses, and what to do to make them better. If you have anxiety about being around several other people, for example, it might help to know that it comes from an evolutionary desire to remain safe. This link has been confirmed by numerous researchers, including Anthony Stevens and John Price, who argued that "psychiatric symptoms are manifestations of ancient adaptive strategies which are no longer necessarily appropriate."[7]

It's important to note that simply *knowing* that your anxiety is founded in your ancestors' completely legitimate fears doesn't cure anything, or even prevent you from feeling it. It could, however, give peace of mind while you are experiencing discomfort.

Our Insignificance

Knowledge about the universe can sometimes have an interesting effect on certain humans. Those who have been to space, for instance, often report an intense feeling of awe regarding how small everything on Earth actually is. Not only do they feel that *we* are small, in comparison to the universe, but that our earthly problems are, as well. This is not news to everyone, of course. Some think we build larger and more complex rockets and telescopes, allowing us a glimpse at farther and more exotic lands, only to reinforce what we had already suspected: *We are not that important.*

This is not to say that each individual isn't important in their

7. Anthony Stevens, & John Price, *Evolutionary Psychiatry: A New Beginning* (2nd ed.) (New York: Routledge/Taylor & Francis Group, 2000).

own unique ways; they are. But humanity as a whole is just one of about 8.7 million species on planet Earth. And the Earth is just one of at least 100 billion planets in the Milky Way galaxy. There are also about 100 billion different galaxies, in addition to ours. This might seem like a depressing statistic at first glance, but it's important to talk about the implications of this.

Understanding how little *humanity* means could also help some people prioritize some of their problems and realize how little other ones actually matter. Of course, this is not true of every concern, but it's hard to worry about getting the wrong number of shots in your morning espresso when you contemplate and understand the scale of our existence. Learning about the complexities of our universe and how it doesn't act with agency may also help us cope with the bad hands we're dealt in life, according to famed evolutionary biologist Richard Dawkins.

"Bad things, like good things, **don't happen any more often than they ought to by chance**," Dawkins wrote in *The Magic of Reality: How We Know What's Really True*. "The universe has no mind, no feelings, and no personality, so **it doesn't do things in order to either hurt or please you**. Bad things happen because things happen."

It is certainly helpful to remember that the universe is not a God, and doesn't do things just to hurt us, but I also understand that this knowledge doesn't come easily and that science can be intimidating for many people. As a result of that, too many of us spend our lives trying to remain ignorant … hoping that ignoring the facts will preserve whatever sense of wonder we cling to. But I don't think ignorance is bliss—I think it just imitates it.

True bliss, I believe, can be found in understanding, learning, and teaching. It can be found in the beauty of a leaf, or in the majesty of a towering mountain. It can be given as freely as a hug, and taken away instantly if we aren't careful. The bottom line is that the better we understand something, the more we can appreciate it. And that applies to our own existence, as well.

6 A Reason to Grow

"Progress in science depends on new techniques, new discoveries and new ideas, probably in that order."

—Sydney Brenner

A common criticism of science in general is that its findings tend to change over time. The discussion usually goes something like this, although the subject matter can change:

> **Science Critic:** "I believe climate change is actually a hoax."
>
> **Science Advocate:** "Numerous studies have shown why that's not true."
>
> **Science Critic:** "Well, science doesn't know everything. Not too long ago, they were saying cigarettes were safe, and now they're saying the opposite. So who really knows?"

I have several problems with this interaction. For one, let's look at the cigarettes claim and the climate change assertion. The fact is that "science" never said cigarettes were safe. What *did* happen, however, is that industry giants used their massive amounts of money to fund "studies" with predetermined results. The tobacco industry was able to save itself for a few years by fighting an intense public relations war using its largely manufactured studies, but in the end, the scientific community was able to restore its good name and establish a strong scientific consensus: cigarettes were actually deadly.

This industry power is one problem with privately funded studies, and it's the same push behind … you guessed it … the "climate change is a hoax" club. "Studies" that show climate change isn't real, or that it isn't caused by humans, are often funded by the oil and gas industries that are reliant upon our current model of energy. So to say you believe climate change is a hoax because science was wrong about cigarettes is … just plain crazy. The logic behind each of these claims directly opposes the logic of the other one.

I chose climate change because of the irony, but I have seen similar scenarios involving numerous topics. In fact, the bigger point has nothing to do with climate change or even cigarettes.

Evolution of Ideas

My real issue with the above situation is that "science," or more accurately, the scientific consensus on any particular issue, is being criticized for merely evolving. For starters, it's not as though scientific revolutions happen every day, shifting the entirety of our knowledge from one end of the spectrum to another on a regular basis. It's a rare occurrence when a study truly challenges what we know about a particular subject and alters the paradigm.

More importantly, though, evolving with new information should be a *good thing*! It's built into the scientific method that old answers should be replaced by better-fitting ones whenever that's applicable, and that should be celebrated, not demeaned. It's what sets science apart from fixed religious belief systems, and what has allowed it to become the dominant system for answering questions about the great unknown.

This evolution of ideas is so successful that I think we should *all* try to incorporate it into our lives. You might think you're good about admitting you're wrong and adopting better ideas, but let's face it, you are biased. We all are. And we are *all* prone to self-justification, which is a process for helping make ourselves feel better about bad choices we have made.

Self-justification is not the same thing as lying or making excuses, according to Carol Tavris and Elliot Aronson, the authors of *Mistakes Were Made (But Not By Me)*:

> Obviously, **people will lie or invent fanciful stories to duck the fury of a lover, parent, or employer; to keep from being sued or sent to prison**; to avoid losing face; to avoid losing a job; to stay in power. But **there is a big difference between what a guilty man says to the public to convince them of something he knows is untrue ('I did not have sex with that woman' ; 'I am not a crook'), and the process of persuading himself that he did a good thing**. In the former situation, he is lying and knows he is lying to save his own skin. In the latter, he is lying to himself. That is why self-justification is more powerful and more dangerous than the explicit lie. It **allows people to convince themselves that what they did was the best thing they could have done**.

That thoughtful book explains how we all make mistakes, and then how we all try to make ourselves feel better about them, but there's an even more important point here. That is, we can all do better, and we can do so by taking a scientific approach to growth.

A Thought Experiment

Think of the last time you learned that you had been wrong about something and corrected your beliefs. If it is difficult for you to come up with an example, you could be a super genius who already knows everything. If that's not the case, and you still struggled to think up a specific instance, then maybe your brain is trying to make it harder for you to remember failures. This is exactly what we talked about above.

That is just one of the obstacles that keeps us from thinking like scientists all the time. The fact is that our brains aren't machines, and they love to take shortcuts often. One of those shortcuts is related to our memories, which change over time and almost always in a way that is more favorable to the person remembering. This phenomenon has been explained by Susan Krauss Whitbourne, PhD, a professor of psychological and brain sciences at the University of Massachusetts Amherst, who used the infamous example of NBC News anchor Brian Williams to put the principles into context.

"When you erroneously recall a past event, **you do so in a way consistent with the way you wish, ideally, to see yourself**. This is what happened to Williams," Kraus Whitbourne explained. "He created a self-defining memory in which he 'remembered' himself as being in a helicopter shot down over Iraq in 2014. In fact, the helicopter was not shot down. As the years went by, **his desire (perhaps unconscious) to see himself as a heroic news reporter in this incident led him to reshape his recall until he apparently believed it to be true**."[1]

It seems like our memories are working against us, keeping us from improving ourselves in order to preserve our own sense of self-worth, and that is before even mentioning the fact that our brains are trained to avoid the uncomfortable feeling of being wrong at all costs. This feeling of having two conflicting ideas is called *cognitive dissonance*, and it's strong enough to make *some* people bury their heads in the sand.[2]

So, if our brains are fighting to be right, how do we actively work on acknowledging we are wrong in order to learn from our mistakes? There may be any number of ways, but for me, it was all about reframing the situation. I had to see being wrong as a positive thing, so that my brain would make more of an effort to remember it, and not in the way one falsely remembers heroically being shot down in a helicopter.

It wasn't an easy task at first. My mind had been trained for decades to see being wrong as a failure, and to forget it. But in the end, I realized that being wrong—and, importantly, learn-

---

1. Susan Krauss Whitbourne, "Why We Remember Things the Way We Want to Remember Them," *Psychology Today*, February 17, 2015, https://www.psychologytoday.com/us/blog/fulfillment-any-age/201502/why-we-remember-things-the-way-we-want-remember-them.
2. It's worth noting that the common stereotype of an ostrich burying its head in the sand is an example of an urban legend itself, as ostriches don't in fact do that. If they did, they wouldn't be able to breathe.

ing from that mistake—really *is* a good thing. After all, *learning* is more of a victory than proving something to a friend or an acquaintance.

I started off by brainstorming times in which I've been wrong in the past and felt some embarrassment, and thought about other, more positive ways to see it. I started to focus on learning new things, growing as a person, and changing my beliefs with evidence.

Because I've retrained my brain in this way, I can think of hundreds of examples of when I was proven wrong, but here are a couple.

For starters, I at one point refused to identify as a "humanist," mostly because I was using out-of-date definitions that left out the importance of non-human animals. I thought humanists had to view humans as inherently superior to other, non-human animals. I was wrong. It turns out that I'm a humanist because I'm unconvinced of supernatural claims, I value critical thinking and evidence over doctrine, and I look to science to help explain the world around us.

In another instance, I once spread false information when I suggested that Homo sapiens "killed off all of our subspecies." Shortly after that, I discovered that my claim wasn't supported by substantial evidence and was, in fact, a minority view among scientists in the field. I thought most experts considered "Neanderthals" a subspecies of Homo sapiens, and I was wrong.[3][4]

---

3. J.J. Hublin, "The Origin of Neandertals," *PNAS* 106, no. 38 (2009): 16022-16027, https://doi.org/10.1073/pnas.0904119106.

4. Katerina Harvati, Stephen R. Frost, Kieran P. McNulty, "Neanderthal Taxonomy Reconsidered: Implications of 3D Primate Models of Intra- and Interspecific Differences," *PNAS* 101, no. 5 (2004): 1147-

None of these are extreme examples, but that's kind of the point. It's not all about shifting every belief you have—it's more about practicing admitting wrongness in everyday situations, and then correcting your own "beliefs" or ideas.

This is all part of using reason and evidence to grow, both intellectually and on a personal level. Believe it or not, this is the same type of growth many critics of science condemn when it comes to the process. After all, if we aren't open to learning new things and improving our ideas, how can we ever answer the important questions that lead to all forms of self-improvement?

1152, https://doi.org/10.1073/pnas.0308085100.

7 Understanding Our Mental Health

"Science can give mankind a better standard of living, better health and a better mental life, if mankind in turn gives science the sympathy and support so essential to its progress."

—Vannevar Bush

Science and reason can help in one really important way: our mental health.

For a long time, mental health was treated as separate from physical health and, as a result, the former received less of a scientific focus than the latter. For instance, let's go back to the 1800s in the United States, when medical science was slow to progress for the first half of the century. But something changed with the Civil War, which prompted a then-unprecedented revolution in how we understand medicine.

In 1862, U.S. Surgeon General Dr. William Alexander Hammond sought to have medical officers send in anything that might help professionals better understand military medicine and surgery. According to one account, that move "had enormous implications on the advancement of the medical field."

"Not only did they create a plethora of resources available for future generations to study, but standardized the practice of diagnosing, treating, monitoring, and if necessary, performing an autopsy for study," according to a public historian. "By drafting these reports and sending specimens, medical personnel was forced to analyze the visual connection between a disease or a wound and its effects on the human body allowing for much more scientific approach to the study of medicine than ever before."[1]

Going Undercover

This revolution in surgical medicine led to huge advancements in how we treat physical wounds, yet twenty years later, there hadn't been the same growth in how we take care of those suffering mental ailments. Enter Nellie Bly, a journalist seeking to expose the U.S. mental "asylum" industry.

Elizabeth Cochrane Seaman, known professionally as Nellie Bly, was a journalist who was assigned an incredible task. In her own words, she was asked to have herself committed "to one of the asylums for the insane in New York, with a view to writing a plain and unvarnished narrative of the treatment of the patients therein." She did exactly that and, after spending 10

1. Paige Gibbons Backus, "Changes in Medicine during the 19th Century," American Battlefield Trust, October 5, 2023, https://www.battlefields.org/learn/articles/changes-medicine-during-19th-century. Backus is a public historian who has been in the field for close to ten years focusing on educational programming and operations at several historic house museums throughout Northern Virginia.

days inside, she published a blistering exposé of the New York City Lunatic Asylum.

"Taking careful notes of both her own experiences and those of her fellow inmates, Bly painted a dire picture in which 16 doctors were assigned to the care of some 1,600 inmates. 'Excepting two,' she recorded, 'I have never seen them pay any attention to the patients,'" according to PBS' report. "She also reported on the cultural insensitivity and language barriers experienced by immigrant women who spoke little or no English and a host of hostile and abusive treatments, from mandatory cold baths to confinement in small, damp, vermin-infested, locked rooms."[2]

Bly helped spur changes at the institution where she was an inmate, but she also caused more funding to be sent to treating mental health on a larger scale. Over time, this moment is one of many that led us to where we are today, at a point in which mental health is largely treated the same as our physical health. This is a more science-based approach, and it has allowed our society to go from locking mentally ill people away to trying our best to treat their illness through therapy and/or medication. This is the approach we should take in our daily lives, too. If you had a severe injury to your foot, for instance, you'd likely seek medical attention. If you have a severe mental infirmity, it makes sense to do the same.

Still, it's important to note that we still need to improve this disparity between mental and physical health. Someone I was

2. Dr. Howard Markel, "How Nellie Bly Went Undercover to Expose Abuse of the Mentally Ill," *PBS*, May 5, 2018, https://www.pbs.org/newshour/nation/how-nellie-bly-went-undercover-to-expose-abuse-of-the-mentally-ill.

once close to asked me for help finding her a psychiatrist, and I made the calls for her, telling doctors that she had psychotic breaks and was a threat to herself and others. Yet nobody could see her for three months or longer. Can you imagine if that happened when someone broke their leg, or if they caught a disease that could kill themselves or others? It's unbelievable but true that, even with all the knowledge our society has at its fingertips, far too often mental health is treated completely differently than physical health, and with more of a stigma, too.

Mental Health Stigma Kills

A big part of the reason physical and mental health are treated differently at times is that the latter suffers from an intense stigma that can shape how society views patients. This stigma is what keeps millions of people quietly suffering instead of seeking the help they need. Danei Edelen of the National Alliance on Mental Illness (NAMI) explains that the stigma actually kills people.

"Mental illness—the topic no one wants to talk about. However, the silence is actually lethal," she wrote, noting that the nation's **suicide rate** is the highest it's been in 30 years. "I have consulted with far too many teens who believe they have a mental health condition, but are afraid to get help. When they confide in me, they often tell me: 'My parents don't believe in mental illness.' Let me say that again: '**My parents don't be-**

lieve in mental illness.'"[3]

Mental health stigma is something most Americans deal with at some point in their lives, and something I personally have dealt with as someone who has been diagnosed with Obsessive Compulsive Disorder (OCD). The truth is that *nobody* is above feeling the effects of mental health stigma. This point is hit home by Dr. Leana S. Wen, an emergency physician and public health professor who regularly spoke "about the stigma of addiction and mental illness and the need to treat mental health with the same compassion and urgency as we treat physical health," when she had an experience of her own.[4]

"Mental health is just as important as physical health, but why is it that we can't find the same access to mental health treatment as we do for physical health ailments? If someone is told they have depression or addiction, why don't we allow them the same compassion and time for healing as if they were diagnosed with high blood pressure or diabetes? These are all illnesses for which treatment exists and recovery is possible," Wen said.

Of her own mental health condition, Wen asked, "Why couldn't I just say that postpartum depression is normal and that I was getting treatment myself?"

Wen talks about how powerful the stigma was, such that it forced her to come up with lies despite being a frequent speaker

---

3. Sabrina Tavernise, "U.S. Suicide Rate Surges to a 30-Year High," *The New York Times*, April 22, 2016, https://www.nytimes.com/2016/04/22/health/us-suicide-rate-surges-to-a-30-year-high.html.
4. Dr. Leana Wen, "I Spoke Out About Mental Illness Stigma. Why Was It So Hard to Admit I Needed Help Myself?" *Elle*, July 27, 2021, https://www.elle.com/life-love/a37133970/i-spoke-out-about-mental-illness-stigma-why-was-it-so-hard-to-admit-i-needed-help-myself/.

on the topic of stigmas in mental health.

"It was clear that my depression was a source of shame. But my shame was the liability, not the actual diagnosis," she wrote.

Treatment Experimentation

When it comes to mental health, the science is *far* from perfect. Anyone who has struggled to find the right combination or type of medications, for instance, can attest to the fact that there is no black-and-white manual to how treatment works. That can be an incredibly painful and, at times, even deadly process.

As I mentioned before, this is something with which I have some level of personal experience. I was diagnosed with OCD in middle school, and it has been a life-long journey of treatment ever since. At the time of writing, I can finally say that I have my symptoms under some level of control, but that is after decades of both therapy and various medications. And unfortunately, that is normal. Under the existing system of mental health care in America, even *finding* the right treatment can be a scientific endeavor filled with trials and pitfalls.

For example, you might start off with the observation that you need help with your mental health. This may stem from your own self-awareness, or perhaps from some comments from your loved ones, but either way, you have now put that thought forward. You might then form a hypothesis about what is wrong with the chemicals in your brain. Perhaps you guess that you have depression, or perhaps it's anxiety you believe is the culprit. To conduct experiments on this, you might try exercise and going outside. If that doesn't cause your symptoms to subside, you could contact mental health professionals, who run diagnostic exercises. Then your doctor would presumably give you therapeutic solutions and/or medications, which is where the data collection comes in. How did you feel after this level of exercise? How about the second day of medication? The third? These are all data points on the scale. Now that you have the crucial information, it's time to draw a conclusion: did the doctor properly diagnose you? Is the treatment working? If not, it might be time to start the entire process over.

This is the reality of the mental health care situation, at least in the U.S. at the time of writing. My hope is that someday this will no longer be the case, and that mental and physical health will be treated as one unit. If that happens, perhaps we could eliminate a lot of the guesswork that goes along with getting treatment these days.

Still, despite the science around mental health not being perfect and the care being subpar in some regions, it's important to follow facts and not be distracted by shiny promises of pseudoscientific "treatments."

The Pseudoscience Problem

One of the main issues with science and mental health is that pseudoscience abounds. Of course, it's true that there are pseudoscientific "solutions" to all that ails you, mental health or no. But there is an especially high number of non-science-based items and services on the market seeking an alternative reality when the mental health system has failed them. This largely unregulated area hasn't gone unnoticed. Take into consideration the work of Scott O. Lilienfeld, PhD, Samuel Candler Dobbs Professor of Psychology at Emory University. He found that, "A careful consideration of errors in thinking is germane to psychiatry and related fields because of the continuing insinuation of pseudoscientific claims into myriad domains of mental health practice."[5]

"Pseudoscientific claims display the superficial trappings of science but lack its substance. As a consequence, they can readily fool nonspecialists—and even specialists, on occasion—into believing that they are well-supported by evidence," Lilienfeld wrote. "In contrast to developed sciences, pseudosciences tend to lack methodological and procedural safeguards against confirmation bias, the deeply entrenched tendency to seek out evidence consistent with one's hypotheses and to deny, dismiss, or distort evidence that is not."

Lilienfeld further states that pseudoscience "poses a serious threat to psychiatric practice and consumers of mental health services."

"Whether they be assertions concerning the use of the Rorschach Inkblot Test to detect anxiety or depression, energy therapies for trauma, gluten-free diets for autism spectrum disorder (ASD), or sensory-motor integration therapy for attention-deficit hyperactivity disorder, numerous mental health claims are contradicted or inadequately supported by research," the study says. "Moreover, many unsubstantiated psychological and psychiatric techniques are widely used. For example, although controlled evidence demonstrates that facilitated communication,

---

5. Scott O. Lilienfeld, "Introduction to Special Section on Pseudoscience in Psychiatry," *The Canadian Journal of Psychiatry* 60, no. 12 (2015):531-3. doi: 10.1177/070674371506001202.

sometimes called supported typing, is ineffective for ASD, survey data indicate that it continues to be widely administered in many quarters, with up to 10 per cent of people with ASD receiving it."

Lilienfeld goes on to list "unsubstantiated techniques" in the history of psychiatry, which include "spinning chairs, tranquilizing chairs, the Utica crib, bloodletting, blistering, purging, leeching, dental removal, and prefrontal lobotomy" and more that "were once believed by many physicians to be effective for psychiatric disorders, but are now recognized as useless or harmful."

"Although the field of psychiatry has learned from many of the painful mistakes of the past, a continued attitude of healthy self-criticism and vigilance will be essential to minimize errors in clinical practice and diminish risk to clients," he writes. "Although controlled data are lacking, some authors have proposed that teaching psychiatry and psychology students about the well-intentioned but disastrous treatment mistakes of previous eras—and the ways in which such mistakes were corrected—may be an effective didactic tool for debiasing them against overconfidence in their clinical judgments."

The solution to this, as the researcher posits, is evidence-based practice, which the American Psychological Association defines in part as "the integration of the best available research with clinical expertise."

"EBP and the accompanying push to develop clinical practice guidelines are crucial steps toward assisting practitioners with the challenging task of sifting pseudoscience from science in psychotherapy and assessment," he wrote.

Self-Care's Connection to Self-Help

A key component of establishing a healthy balance when it comes to mental health is self-care.

When you hear the term "self-care," you might think of the things you do to relax, such as taking a bath, but the World Health Organization defines self-care as "the ability of individuals, families and communities to promote health, prevent disease, maintain health, and cope with illness and disability with or without the support of a health worker." Both definitions are important as we see how self-care can help us achieve more.

It's a scientific fact that self-care can improve productivity. According to Dr. Matt Glowiak, who has a PhD in counselor education and supervision, "By taking some time out to engage in this practice [of self-care], you may relieve the pressures of everyday life and reset yourself to get back to a healthy point where you can be more productive again."[6]

"Considering the costs associated with mental health services, lost wages and more, spending some time on yourself may ultimately benefit everyone," according to Dr. Glowiak. "Burning the candle at both ends, so-to-speak, comes with significant consequences, which may include but are not limited to burnout, depression, anxiety, resentment and a whole host of other negative implications."

Dr. Glowiak further noted that having a self-care routine "has been clinically proven to reduce or eliminate anxiety and

---

6. Dr. Matt Glowiak, "What Is Self-Care and Why Is It Important for You?," *SNHU*, January 23, 2024, https://www.snhu.edu/about-us/newsroom/health/what-is-self-care.

depression, reduce stress, improve concentration, minimize frustration and anger, increase happiness, improve energy and more."

"From a physical health perspective, it has also been clinically proven to reduce heart disease, stroke and cancer," he wrote, before elaborating about the benefits of self-care when it comes to increasing productivity.

"Taking care of yourself allows you to recharge and rejuvenate, leading to increased productivity, improved concentration and better problem-solving abilities. When you prioritize your well-being, you have more energy and mental clarity to tackle daily tasks," according to the expert.

Now, it is true that self-care makes you more productive and can therefore be a key component of success in whatever field or task to which you are dedicating yourself. But it's equally true that self-care shouldn't necessarily be about helping you be productive—it can also be a reminder that you're more important than accomplishing goals.

8 The Science of Emotions

"Everything we do is for the purpose of altering consciousness. We form friendships so that we can feel love and avoid loneliness .... We read for the pleasure of thinking another person's thoughts. Every waking moment, and even in our dreams, we struggle to direct the flow of sensation, emotion, and cognition towards states of consciousness that we value."

—Sam Harris

One of the few situations when *feelings* are thought to count more than facts is when matters of the heart are concerned. This is why people are often encouraged to "follow their heart" when it comes to romance.

But does that mean that we should *only* go with our gut on these important issues? I would argue that facts, evidence, and—to some extent—science, also have a role. In fact, I believe that, even when you are listening to your emotions, you could still be following science.

That's because emotions are not some mystical thing beyond reach; they are directly rooted in science. Given the right technology, emotions are even measurable when it comes to analyzing brain waves.

Bryn Farnsworth, scientific program manager at Lundbeck Foundation, argues that the "physiological aspect of emotions is not only central to their experience but also to their measurement."[1]

"Advances in technology have enabled researchers to measure emotions through various methods: Electroencephalography (EEG) captures brain activity, Functional Magnetic Resonance Imaging (fMRI) visualizes blood flow changes in response to emotional stimuli, and Electrocardiography (ECG) monitors heart rate variations," Farnsworth writes. "These objective measures provide a window into the automatic, instinctual nature of emotional responses, underscoring the difference between the physiological experience of emotions and the subjective experience of feelings."

Emotions are nothing but chemical reactions and the limbic system of your brain, and they can be measured with the right tools, yet it goes beyond that. They are also based on how we *feel*, our past experiences, and more. Positive and negative emotions both exist, so the question is, how do we most efficiently enhance the former while limiting the latter?

Unhealthy Coping Mechanisms

Therapy is one of the best things you can do to help manage these very real emotions. But as we discussed in the previous

---

1. Bryn Farnsworth, "How to Measure Emotions and Feelings (And the Difference Between Them)," Imotions, https://imotions.com/blog/learning/best-practice/difference-feelings-emotions/.

chapter, finding the *right* type of mental health care can be a chore in the U.S. So it shouldn't come as a surprise that lots of people either don't try or don't find what they need in therapy, leading them to resort to unhealthy mechanisms for coping.

Religion can even be something people turn to when they're down and in search of meaning or solace. Personally, I don't feel the need to be comforted by preachers or deities. Instead, I face my troubles by grounding them in reality and finding out what to do to get past them. But there *are* people who report that religion helps them.

While some turn to faith, others may dive into substances. Addictions, whether they are to alcohol, drugs, or anything else, often stem from this attempt to self-medicate or to numb certain emotions. Licensed psychologist Dr. Glenn Patrick Doyle has spoken about this particular issue.

According to Doyle, "You don't just 'treat addiction.' You end up treating anxiety, depression, PTSD, loneliness, rage, despair, toxic secrets, regret, undiagnosed head trauma, untreated ADHD. Then you realize addiction is often someone's best attempt to cope when they don't see other options."

If you're unhappy and seeking out unhealthy coping mechanisms, social scientist Dr. Arthur C. Brooks may have a better way. During a presentation hosted by the Butler Institute for Free Enterprise Through Entrepreneurship, Brooks, who has written for *The Atlantic* about happiness and loneliness, compared happiness to a traditional Thanksgiving meal of pumpkin pie, mashed potatoes, and turkey.[2]

2. Hillary Chabot, "Why Entrepreneurship Is Key to Lifelong Happiness," *Babson Thought & Action*, April 21, 2022, https://entrepreneurship.babson.edu/arthur-brooks-happiness.

"Now, why do I bring that up? Because that's a good metaphor for happiness. All foods are made up of three micronutrients—fat, carbohydrates, and protein," he said. "Happiness requires three macronutrients in abundance and balance. They are enjoyment, satisfaction, and purpose."

Searching for Answers

Let's take a look at *one specific* feeling, loneliness, as an example of when science doesn't have all the answers (but is looking). Loneliness is a negative feeling, and, unfortunately, it has become chronic for many people over the last few generations. To search for solutions, researchers embarked upon a study called RECETAS (Re-imagining Environments for Connection and Engagement: Testing Actions for Social Prescribing in Natural Spaces).

That study, according to Science Magazine, "tries to create an environment where deep connections can emerge."[3]

"Among studies focused on alleviating loneliness, few have had the size or scope of RECETAS, which will recruit more than 1000 participants from diverse social and economic groups across six countries in Europe, South America, and Oceania," according to the outlet's analysis. "The study has €5 million in funding from the European Union and another AU$400,000 from Australia's National Health and Medical Research Council."

3. Kelly Servick, "Striving to Connect," *Science*, April 25, 2024, https://www.science.org/content/article/can-science-find-ways-ease-loneliness.

Cognitive behavioral therapy (CBT) is one of the psychological interventions that has shown some success with loneliness. The Science article claims that, "An evidence review released in September 2023 by the U.K. government judged psychological approaches to be the only category of loneliness interventions to have so far demonstrated a 'large statistically significant effect' on loneliness."[4]

"But because one-on-one sessions with a trained therapist tend to be costly and time-intensive, some researchers are trying to capture their essence with online programs. [Researchers] found significant reductions in loneliness in two randomized trials of their online CBT intervention where participants complete assignments and correspond with a therapist who offers weekly feedback on their work," writes Kelly Servick.

Science can also point us in another, much simpler direction when it comes to loneliness. World-renowned family therapist Virginia Satir said, "We need 4 hugs a day for survival. We need 8 hugs a day for maintenance. We need 12 hugs a day for growth."

Christine Comaford, a business strategist who focuses on leveraging neuroscience to "create remarkable leadership," took that even further. According to Comaford, "Hugs boost oxytocin levels, which heal feelings of loneliness, isolation, and anger."[5]

---

4. Margherita Musella, Dr. Joanna Blodgett, and Dr. Fran Harkness, "Loneliness Interventions across the Life-Course: A Rapid Systematic Review," 2023, https://assets.publishing.service.gov.uk/media/64f1f88fa78c5f000dc6f455/Loneliness_interventions_across_the_life_Rapid_Review.pdf.

5. Christine Comaford, "What Stress, Change and Isolation Do to Your Brain," *Forbes*, October 20, 2018, https://www.forbes.com/sites/christinecomaford/2018/10/20/what-stress-change-and-isolation-

"Extended hugging (20 seconds or more) boosts your serotonin levels, causing you to feel happy and more positive emotions overall. When you see another person look at you with kindness and affection you remember that you're loved, just like you did as a child when your parents looked at you and acknowledged you were here and they were happy you were," she wrote, adding, "A hug results in some pressure on your sternum, which then stimulates the thymus gland, which then regulates and balances the body's production of white blood cells, which keeps you healthy."[6]

While the science may not be fully in on every detail of how loneliness works or how to permanently cure it, that doesn't mean we can't use the data we *do* have to make decisions that positively impact our mindset. Take this finding about hugs, for instance. As researchers continue to search for answers about the deeper questions associated with loneliness, we can use this information to help improve our day-to-day lives. In addition to finding someone in your life who can be a "Hug Buddy" and hugging your pet, Comaford recommends readers to hug themselves.

"This may sound silly but it works. Remember the goal is oxytocin and serotonin release, so you may want to turbo-charge this experience by listing all the things you love about yourself," she wrote.

---

do-to-your-brain.

6. Christine Comaford, "Are You Getting Enough Hugs?" *Forbes*, August 22, 2020, https://www.forbes.com/sites/christinecomaford/2020/08/22/are-you-getting-enough-hugs.

Scientific Fear

Fear is a troublesome emotion, but not one that is completely immune from the powers of science. Enter Margee Kerr, a sociologist at the University of Pittsburgh who also happens to run a haunted house program that she uses to terrify people and then study their fear. Kerr and Greg Siegle, Pitt professor of psychiatry and associate professor of clinical and translational science, analyzed data from the experiment to find the positive in what scares us all.

"The prevailing view in psychiatry is that fear is a big part of a lot of disorders, so it gets a bad rap," Siegle said, according to the university's official press release. "There really aren't a lot of spokespeople for the good side of fear, except the people who say that fear prevents us from doing bad things. Margee is fairly unique in that she's speaking for the upsides of fear: that it's exhilarating, that it's fun. And that fundamentally changes our relationship to our own emotional state."[7]

Siegle also added, "As long as you treat fear as inherently bad, I think people won't want to engage with it, which makes people avoid fearful stimuli. And a lot of anxiety disorders are about avoidance … We're all going to have to live with fear, and if we understand that it might have some upsides, we might look at our experience differently."

It turns out that you can even use science to calm yourself during flight turbulence. Gordon Malloy, Helmsman and Lieutenant aboard the fictional USS Orville, had some advice on

7. "Pitt Researchers Are Examining Why Fear Can Be Fun," *Pittwire*, October 27, 2017, https://www.pitt.edu/pittwire/features-articles/researchers-examine-fun.

that front. In trying to calm someone who is scared to fly because of turbulence, Malloy explained the physics behind the phenomenon, saying, "That's never going to hurt you."

"Here's why the physics are in your favor. Picture the plane suspended in a giant mound of JELL-O. When there's turbulence, it's almost like someone is shaking the JELL-O plate. Now, you're gonna get bounced around, but other than that, you're not going anywhere. You're going to stay right where you are, inside that JELL-O … so next time you're on a bumpy flight, just picture jiggling JELL-O and knock back five or six cocktails and you'll be fine."

9 Evolution of Our Ethics

"The first step in the evolution of ethics is a sense ofsolidarity with other human beings."

—Albert Schweitzer

For a lot of people, part of their solution for improving themselves is to be more empathetic toward other individuals. Being good to others is one of the keys to a happy life, according to psychologist David Cregg. He argues that doing positive things for people can improve our sense of social cohesion as well as our purpose in life, and can even support us in living extended and healthier lives.

"Social connection is one of the ingredients of life most strongly associated with well-being. Performing acts of kindness seems to be one of the best ways to promote those connections," Cregg said, according to Ohio State News. "Something as simple as helping other people can go above and beyond other treatments in helping heal people with depression and anxiety."[1]

1. Jeff Grabmeier, "Feeling Depressed? Performing Acts of Kindness May Help," *Ohio State News*, January 10, 2023, https://news.osu.edu/feeling-depressed-performing-acts-of-kindness-may-help.

This is something that's backed up by research from around the globe, according to the UK charity the Mental Health Foundation.

"Those of us who are kind and compassionate experience clear benefits to our well-being and happiness. We may even live longer. Kindness also helps reduce stress and improve our emotional well-being," the foundation says. "Acts of kindness have the potential to make the world a happier place."[2]

One study, called "Happy to help? A systematic review and meta-analysis of the effects of performing acts of kindness on the well-being of the actor," provides a meta-analysis of the effects of helping on the happiness of the one doing the helping.

"Helping others makes you happy, but the effect is relatively modest. Further empirical work testing the implications of more specific theories of social, cooperative and altruistic behaviour is needed to determine whether the effect might be larger for some types of helpers, when helping some types of recipients," the study's authors wrote. "This research will advance our understanding of the causes and consequences of kindness, and help practitioners to maximise the effectiveness of kindness interventions."[3]

A lot of modern self-help advice has taken on a self-centered tone, basically suggesting that readers should just "do you"

---

2. "Kindness Matters Guide," Mental Health Foundation, https://www.mentalhealth.org.uk/explore-mental-health/kindness/kindness-matters-guide.
3. Oliver Scott Curry, Lee A. Rowland, Caspar J. Van Lissa, Sally Zlotowitz, John McAlaney, Harvey Whitehouse, "Happy to Help? A Systematic Review and Meta-Analysis of the Effects of Performing Acts of Kindness on the Well-Being of the Actor," *Journal of Experimental Social Psychology* 76, (2018): 320–329.

and ignore others. But I feel like this take, while perhaps helping people avoid certain negative interactions in the short term, doesn't do as much to help an individual grow and mature. Instead, research appears to suggest that helping others helps ourselves. But it's also important to take care of yourself, so balance is crucial.

Being good to others is something that comes naturally to us. Don't get me wrong, so does selfishly undercutting our competition. But it's in our DNA to work with those in our tribal affiliation, and that's something many groups, including certain religious fundamentalists, tend to ignore. As opposed to seeing being nice as a type of social care, these individuals see it as something bestowed upon humanity by an all-powerful deity.

Depending on who you are, being more empathetic and caring might mean something different. Maybe you just need to start out doing one kind act per day for a stranger. Maybe you require a little more than that. But regardless of where you are on that scale, understanding the origins of empathy could help you carry out that day-to-day plan.

Origins of Empathy

Most believers understand that being good goes deeper than one's religious belief system, but, unless you've been living under a rock, you've likely heard a religious apologist claim that you can't be "moral" without religion … or, more specifically, their religion.

Some fundamentalist believers in the Abrahamic traditions often make this argument despite the fact that their holy books contain many verses that appear to bestow divine justification on actions the average person today would agree are anything but "moral." In fact, in many cases, the Bible and the Qur'an both sanction behavior that is considered downright unethical.[4]

From the perspective of a religious studies graduate, the amoral prescriptions in ancient holy books make total sense. These are actions that were considered "holy" or "sacred" thousands of years ago, and they are preserved in time through oral and written accounts. Today, they are in many cases completely irrelevant in today's modern, interconnected society.

So, if morality doesn't actually come from these religious traditions, then where does it come from? In *Hi, I'm an Atheist*, I talk about how we can see how moral values could have evolved over time within society in conjunction with the evolution of other social constructs, such as a sense of community, nationalism, and even democracy. Here's an excerpt:

"I think that what we now call 'morality' has evolved—as nearly all social and physical human attributes have—to aid us in survival and, ultimately, reproduction. **This morality requires that we be guided only by a conscience (or 'moral sense') and not by a god or gods.**

The evolution of morality can be thought about in similar terms to physical evolution. We can determine which aspects of our ancestors' behaviors best allowed for the primitive communities to flourish—**therefore ensuring that those behaviors flourished in future generations and remain present (and continually changing) to this day**."[5]

---

4. See more examples in my previous book, *Disproving Christianity*.
5. David G. McAfee, *Hi, I'm an Atheist* (New York: St. Martin's Publishing Group, 2021).

So, next time you hear someone say that you can't have morality without religion, be sure to remind them that ethics don't come from dogma, just like trees don't come from houses. Moral lessons were added to holy books to keep order in societies that didn't have complex justice systems.

Putting It to Use

Knowing about the importance of helping others, both for society as well as for our own individual fulfillment, is one thing. Choosing to take the steps necessary to increase your charity work and otherwise improve your empathetic actions is an entirely different task.

A famous Harvard study, the Study of Adult Development, included 75 years of research on participants and their children. Its intention was to find out how the various lives developed over the years. Dr. Robert Waldinger, the current director of the program at the time of writing, explained what they learned from the massive project.

"What are the lessons that come from the tens of thousands of pages of information that we've generated on these lives?" Waldinger says in his TED Talk. "Well, the lessons aren't about wealth or fame or working harder and harder. The clearest message that we get from this 75-year study is this: Good relationships keep us happier and healthier. Period."[6]

6. Robert Waldinger, "What Makes A Good Life? Lessons from the Longest Study on Happiness," TED Talk, transcript: https://singjupost.com/wp-content/uploads/2020/03/What-Makes-A-Good-Life_-Lessons-From-the-Longest-Study-On-Happiness-by-Robert-

Waldinger added, "It turns out that people who are more socially connected to family, to friends, to community, are happier, they're physically healthier, and they live longer than people who are less well connected."

So, if we take this extremely thorough research project into consideration, it seems a key focus on improving our lives *could* actually be found in helping those around us and fostering our closest connections. To do this, you'll have to practice things like active listening, empathetic thinking and actions, and being there for those in your life.

One potential source for information on this subject is Greater Good Magazine, which provides "science-based insights for a meaningful life." The magazine is published by Greater Good Science Center (GGSC) at the University of California, Berkeley, and it seeks to "explore the roots of happy and compassionate individuals, strong social bonds, and altruistic behavior—the science of a meaningful life."

Greater Good says social connections are also referred to as "social capital" for a good reason:

"They are a valuable resource in life, creating moments of positivity and fun, supporting us through good times and bad, and exposing us to new ideas and new people. If we cultivate healthy connections, the research promises a longer, happier, and more prosperous life."

As far as tips for making this happen, Greater Good recommends showing interest in others' good news, appreciating your loved ones more by imagining your life without them, and having conversations and spending quality time with those you care

---

Waldinger.pdf.

about. It also suggests you can meditate with someone, listen to music together, or even figure out your own attachment style.

"If you have commitment, trust, or attachment issues, you could significantly improve your relationships by understanding your attachment style and trying to develop a 'secure' one," according to the magazine. "Social connections are tied up with other keys to well-being. Cultivating gratitude, empathy, altruism, compassion, and forgiveness will not only enhance your personal well-being but also benefit your relationships at work and at home."[7]

7. "What Is Social Connection," *Greater Good Magazine*, https://greatergood.berkeley.edu/topic/social_connection/definition#how-cultivate-social-connection.

10 Not All Scientists Wear Lab Coats

"I hope that someday scientists can be considered heroes again."

—Walter Isaacson

The only reason we can take scientific findings and apply them to our lives is because of dedicated researchers and professionals who actually make those discoveries. Think about your day-to-day life, and all the activities that are only possible due to various scientific advancements. Not just your computer, TV, tablet, etc., but even the fact that you have access to indoor plumbing and clean drinking water is due to the work of historical scientists. Yet in society, scientists aren't really revered, or even respected, as they should be.

Most Americans know LeBron James, Travis Kelce, and Serena Williams, but it's rare that a modern scientific thinker has achieved any sort of notoriety, let alone that sort of star status that sports heroes get. Consider the rare exception that is Katie Bouman, a 29-year-old computer scientist who created an algorithm that made possible a jaw-dropping black hole photo and became an internet meme overnight after her celebratory photo went viral. Her photo generated sexist hatred coming from some corners of the internet, but a male colleague named Andrew Chael defended Bouman when others tried to discredit her by noting that Chael is the primary developer of the imaging software library.

"While I wrote much of the code for one of these pipelines, Katie was a huge contributor to the software; it would have never worked without her contributions and the work of many others who wrote code, debugged, and figured out how to use the code on challenging EHT data," he said.[1]

History Ignores Great Minds

Similarly, when we look at history, it's not uncommon for major, groundbreaking scientists to go relatively unnoticed. Mathematician and astronomer Ibn al-Haytham, for example, isn't a household name in the U.S. Yet he is the one who laid the groundwork for the scientific method itself.

"Born in the mid-10th century in what is now Iraq, Ibn al-Haytham, known to English speakers as Alhazen, was a man of endless curiosity. At a time when the Arabic-speaking world was the epicenter of scientific inquiry, Alhazen was one of its brightest stars," writes Gemma Tarlach for Discover Magazine.[2]

---

1. Kathleen Joyce, "Internet Trolls Attempted to Discredit Katie Bouman's Work on Black Hole Project," *Fox News*, April 13, 2019, https://www.foxnews.com/science/internet-trolls-discredit-katie-boumans-black-hole-project.
2. Gemma Tarlach, "The Unsung Heroes of Science," *Discover*, October 19, 2017, https://www.discovermagazine.com/the-sciences/the-unsung-heroes-of-science.

"He wrote more than 100 books on physics, mathematics, and astronomy, among other fields, and is believed to be the first to explain how our brains create the illusion of the moon appearing larger near the horizon. His pioneering work on optics inspired the likes of Roger Bacon and Johannes Kepler centuries later. But Alhazen's creation of the scientific method is his most far-reaching achievement."

Thinking of overlooked scientists always reminds me of that old joke, "What was the greatest thing James Watson and Francis Crick discovered? Rosalind Franklin's notebook." Her contributions to the discovery of the molecular structure of DNA have been ignored for too long.

To state it simply, Franklin helped us better understand more about ourselves by improving our understanding of our DNA. Franklin found out more about DNA with a picture that took more than 100 hours to create. It's called Photo 51, and it's the first picture ever of DNA.[3]

Sadly, her discovery has been largely overlooked. Thankfully, the times are changing, and Franklin can get the credit she so rightfully deserves. After all, without her work in science, none of us would know as much about our inner workings. By focusing on the small side of life, she changed the whole entire world.

Changing Our Perspective

Having a paradigm shift that changes how we see scientists is a

3. To see Photo 51, read "The Story behind Photograph 51" by Professor of Molecular Biophysics Brian Sutton: https://www.kcl.ac.uk/the-story-behind-photograph-51.

pretty far-fetched goal. But a more realistic approach would be to work individually to look at more scientists and thinkers as our role models. After all, emulating the top thinkers is bound to improve anyone's life.

For example, Charles Darwin happens to be one of the most influential scientists of all recorded history. He helped uncover how evolution works, positing that it was by natural selection, and challenged the thinking of his time. He can also teach us a lot about how to keep persevering, because he almost gave up, but he never did.

In 1831, Darwin was invited on a trip around the world to work as a naturalist on the **HMS Beagle**. This trip ultimately shaped Darwin's future and our own understanding of how life evolves on Earth, but it almost didn't happen because Darwin's father wasn't a fan of the idea.

Darwin was incredibly intrigued by the invitation, but he was almost convinced by his father's reasoning on the matter. Instead of giving up, though, Darwin sought out his uncle **Josiah Wedgwood**, whom his father respected greatly, and the two were able to change Darwin's father's mind, according to the American Museum of Natural History.[4]

What we can learn from this historical example is that sometimes it's worth following your dreams, even when others don't see your vision. Sometimes, great things will happen as a result.

The fact is that, without any supernatural or divine intervention to make your choices, you are the author of your own life, and that means you get to choose how your story is written.

---

4. "A Stunning Invitation," American Museum of Natural History, https://www.amnh.org/exhibitions/darwin/a-trip-around-the-world/a-stunning-invitation.

It also means that you can overcome writer's block if you just keep scribbling, putting down words on the page. More often than not, you'll come up with something beautiful.

Next time you feel like giving up on something important to you, remember that the best things in life don't come overnight. Sometimes, greatness takes some time. If you persevere, taking things one day at a time and making little choices that work toward your goal, you will eventually get there. And when you arrive, you might be surprised by how impressive the view is, and proud of the journey you took.

It's not always easy to keep going, but even if that's the *only* thing you can do, sometimes it's enough.

Evolution as a System of Betterment

Darwin was a revolutionary, and there are multiple ways in which one can draw inspiration from him and his work. Aside from the perseverance Darwin showed as a scientist and a thinker, we can also look at his work itself for guidance.

Darwin's theory of evolution by natural selection *did* change biology, and even science in general, but it did even more than that. It changed the entire paradigm for how we look at change in general. Yes, Darwin was writing about the animal kingdom and how we all relate to one another, but *evolution* means much more than that.

We can evolve as people, and our ideas can evolve, but it's even bigger than that, too. It's the actual process of natural selection that has now become relevant to basically every single facet of our society. It doesn't only apply to animals and how we change over time; it also works with thoughts, ideas, and beliefs.

One way we can use evolution by natural selection as a means to better our own lives is to look at it as a mechanism for honing what we are putting out into the world. For example, let's imagine one of your long-term goals is to be more well-spoken or to learn a new skill. The best way to accomplish either of those tasks might be to follow the same pattern as natural selection does.

This is something that rapper and playwright Baba Brinkman has addressed in his work. In a song entitled "*Performance. Feedback. Revision,*" Brinkman suggests natural selection is the same process by which he writes his shows and by which human beings "*ever learn to do anything.*"

> And evolution is really just kind of an algorithm that goes
> Like this: Performance. Feedback. Revision
> So the genetic code of every living thing was written
> Like this: Performance. Feedback. Revision
> So the genes are like a text with a thousand pages
> And revisions occur in the random changes that come from mutations.
> And when they see the light, well, that's the performance, that's the phenotype.

In a follow-up evolution song, Brinkman says, "Performance, feedback, revision, that's how I mastered this rhythm."

> It took a bit of practice, but I cracked it now. I rap with mad precision
> Performance, feedback, revision, that's art and architecture
> And engineering, that's how we peer into the dark with guesswork

> Performance, feedback, revision, that's the scientific method
> You put a hypothesis to a double-blind test, and if it isn't right, then reject it
> Performance, feedback, revision, that's how your brain is wired
> Behaviors that fire together can bind together and change desires
> Performance, feedback, revision, that's your immune system
> Antibodies react to pathogens and reproduce when a new one gets in
> Performance, feedback, revision, that's natural selection
> Some genes die, and others survive, and which is which is fitness
> Performance, feedback, revision, this isn't just music
> It's descent with modification, and you owe your whole existence to it.

The beautiful part about this process is it's applicable to virtually everything, and it can improve anything. This is how Darwin left behind more than just a scientific theory, but also a mechanism by which we can get better at anything.

Therapeutic Techniques for Improving Your Life

Evolution by natural selection isn't the only scientific principle we can use to make our daily lives better. Cognitive behavioral therapy is yet another one. If you're unfamiliar with this concept, it's commonly used as a therapeutic tool, especially for those patients who may have a mental illness. But CBT techniques can be isolated from the therapy and used to help people function properly day-to-day, according to the Mayo Clinic.

"Cognitive behavioral therapy (CBT) is a common type of talk therapy (psychotherapy). You work with a mental health counselor (psychotherapist or therapist) in a structured way, attending a limited number of sessions. CBT helps you become aware of inaccurate or negative thinking so you can view challenging situations more clearly and respond to them in a more effective way," the Clinic states. "CBT can be a very helpful tool — either alone or in combination with other therapies — in treating mental health disorders, such as depression, post-traumatic stress disorder (PTSD) or an eating disorder. But not everyone who benefits from CBT has a mental health condition. CBT can be an effective tool to help anyone learn how to better manage stressful life situations."[5]

Some methods for utilizing these therapeutic techniques in daily life include focusing on identifying and challenging negative thoughts, practicing mindfulness (also known as meditation), tracking your thoughts and emotions via a journal, breaking down large tasks into smaller, more manageable steps, scheduling enjoyable activities, and even slowly exposing yourself to situations that cause anxiety while incorporating breathing exercises.

One key way to implement some of these practices could be the public school system. Diana Joyce-Beaulieu, PhD, NCSP, and Michael L. Sulkowski, PhD, NCSP, have written about this potential.

Together, they developed *Cognitive Behavioral Therapy in K-12 School Settings: A Practitioner's Toolkit*, in which they "provide prac-

5. "Cognitive behavioral therapy," *Mayo Clinic*, https://www.mayoclinic.org/tests-procedures/cognitive-behavioral-therapy/about/pac-20384610.

titioners with an easily accessible and practical guide for implementing basic cognitive behavioral counseling strategies in K-12 school settings."[6]

"Because of the unmet mental health needs displayed by millions of students in these settings and the advancements in the training and provision of school mental health services during the past couple of decades, school-based mental health professionals, such as counselors, school psychologists, social workers, and others, are increasingly being asked to provide evidence-based counseling and intervention services such as CBT," the authors wrote.

Anyone Can Be a Scientist

We must understand and acknowledge the importance of scientists and the theories they leave us, as well as groundbreaking research in general, but there's an even bigger point here: We can all live more scientifically.

For starters, it's important to encourage people who have the desire to do so to follow a life of science. Kids should know that science is a viable career option, if they have the drive and curiosity to pursue it. And adults should know that it's never too late to start learning more about science.

One inspiring example to note here is Elisa Quintana, an astrophysicist at the Goddard Space Flight Center. She was also a key part of the discovery of the first Earth-sized planet outside of our solar system. Yet her path didn't always look like it was

6. Diana Joyce-Beaulieu and Michael L. Sulkowski, *Cognitive Behavioral Therapy in K-12 School Settings: A Practitioner's Toolkit* (New Jersey, USA: Springer Publishing Company, 2015).

heading toward space.

Not only was Quintana *not* a stargazer as a young kid, as many of her colleagues were when they were children, but she also didn't do well in math or science in school. She went to a local community college, as opposed to a major, well-known university.

"I went through a tween phase where I wasn't interested in anything in school," Quintana said, according to NASA's biography. "I went to junior college because it was just the thing to do. No one went to grad school and no one in my family was a scientist. It just wasn't a thing."

She added, "I was definitely a late bloomer … I think it's important for people to know that there isn't just one path."

Part of what changed things for Quintana was what we've talked about in this very chapter: she found a scientific role model. Specifically, she transferred to the University of California, San Diego, and got a former astronaut (Sally Ride) as an adviser.

"I was starstruck," Quintana said. "I went from having no role models in science to having the ultimate role model: America's first woman in space. I remember always admiring how strong and confident she was."

That pushed Quintana to pursue physics, and then become a research scientist with the SETI Institute, and later a NASA senior postdoctoral fellow at Ames Research Center. She later led the team that discovered Kepler-186f, which NASA says is "an Earth-sized exoplanet in the habitable zone of a red dwarf star."[7]

7. "Elisa Quintana: Astrophysicist—Goddard Space Flight Center," NASA, https://science.nasa.gov/people/elisa-quintana/.

But this goes beyond being a late bloomer in science. Most people who read these words will never do anything in the world of science, just based on statistics alone, and that is okay. But stories like Quintana's mean much more than that. They help us see that passion and a desire to learn can propel us in any direction. Anyone in the world can be more of a scientist in their daily life, yes, but it's also never too late to follow any other passions, including those that have been lurking beneath the surface of your self-consciousness for many years.

Is there something in the arts that you wanted to do but weren't able to? Maybe now it's time to explore it once again. But keep in mind to do it rationally, by practicing in your spare time and seeing if you have the potential to sustain yourself with your craft.

11 Learning from Nature

"People talk about caterpillars becoming butterflies as though they just go into a cocoon, slap on wings, and are good to go. Caterpillars have to dissolve into a disgusting pile of goo to become butterflies. So if you're a mess wrapped up in blankets right now, keep going."

—Jennifer Wright

Many years ago, I was giving a presentation in which I made a remark about the scientific efficacy, or lack thereof, of astrology. A woman by the name of Kathleen approached me and said, "*I still have a hard time letting go of this one. I liked identifying with a lion.*"

I replied, "*You can still identify with a lion if you don't believe in astrology. I personally like snakes.*"

This may sound simple—and it is—but it's also important because it demonstrates the power we have over who we are and from where we draw our inspirations. For many people, one of the best places to get inspiration in life is nature—after all, we are a part of it. This statement that humans are part of nature isn't a philosophical statement; it's a fact. We are animals who have evolved to fit our needs just like all the others.

Humanity's scientific pursuits, beginning thousands of years before the Common Era, have revealed countless characteristics of animals that have survived for generations in a dangerous world, so it would be irresponsible to ignore that information when it comes to improving our lives. It's also important to note that these inspirations can come from surprising sources. Richard Dawkins, for instance, famously said, "If lions were discovered weaving antelope-catching nets ten lion-lengths wide, it would be headline news. Yet spiders weave intricate insect-catching nets, hugely bigger than themselves, and we treat them as commonplace. What's the difference?" Maybe it's worth looking at what all animals, including the humble spider, can teach us.

If you look hard enough, you can find motivation to make yourself better just about anywhere. Since the beginning of time, one of the most common sources for this type of self-betterment has been animals. The lion is courageous, the snake is cunning, and the owl is wise. Our ancestors used these archetypes as labels, and many people still do. In fact, it could be argued that the astrological system that roughly 25–30% of Americans believe in, still retains some of these practices.

The Irwin Test

One person who built a legacy from taking inspiration from nature is the late conservationist Steve Irwin. He truly believed in the importance of every little creature in the animal kingdom, and he considered their preservation to be paramount.

"I believe that the time has come where if we don't get animals into people's hearts, they're going to go extinct," he once told the late Larry King of CNN. "We're running out of time right now, Larry, right now as we stand here."

Irwin is also known for saying, "We don't own the planet Earth; we belong to it. And we must share it with our wildlife."

Irwin's desire to protect wildlife manifested itself in what can only be described as a friendly and loving demeanor toward the animals themselves. Another popular Irwin quote is, "I have no fear of losing my life—if I have to save a koala or a crocodile or a kangaroo or a snake, mate, I will save it."

Known for his excitement when he spotted a cool critter in the wild, Irwin would often compliment the animals he encountered. Even when handling the most venomous snake in the world, Irwin was kind and loving to the creature as it tried to aim its deadly bite at his face. "I know you are, you're a beautiful snake, you're absolutely gorgeous."

"Beautiful snake!" he shouts again as the "grumpy" animal goes for the kill.

"A lot of people think snakes are evil, ugly monsters that cruise around killing people … What a beautiful snake. She knows I mean her no harm … What an honor to share space with such a beautiful animal," Irwin said. "I'd rather deal with this snake than a lot of people I know. I reckon that people are a lot more dangerous."

It is this positive, sweet, and empathetic attitude toward the animal kingdom that leads us to the Irwin test, which is a popular internet meme and phenomenon. It's a simple-but-effective exercise that involves a little bit of imagination, and it goes like this:

If you are feeling down on yourself, and your internal criticisms are mounting, you can employ the Irwin test by simply imagining you are a little lizard or snake just gently strolling through tall blades of grass. You're in an expansive field with everything you need right nearby—you have access to yummy foods and running waters. You can feel the sun heating up your body, as well as the soft, slightly wet dirt beneath your feet as you move. Just then, Irwin's smiling face appears, and he tells you how beautiful and amazing and misunderstood you are. He showers you with compliments because he truly does love you and all animals. Imagine all the sweet things he might say to that little creature, who is you, and how it would make you feel. And *that* is the Irwin Test.

Nature-Based Scientific Advancements

At the intersection of nature, science, and human advancement, we have one stand-out issue, and that's *longevity*. For years, scientists have been taking their marching orders from nature by researching how some living beings are able to live much longer than we do.

"What are the mechanisms underlying great longevity and can we apply such knowledge to enhance the health and longevity of humans?" asked researchers associated with the National Institute on Aging. The study, called *Aging and longevity in the simplest animals and the quest for immortality*, concludes that "we may be able to derive benefits to humans from research on phenomena such as dedifferentiation and regeneration in the simplest animals."[1]

1. Ronald S. Petralia, Mark P. Mattson, and Pamela J. Yao, "Aging

"[T]hese studies may provide valuable information useful to overcome human diseases and other problems of aging. Stem cell research already has shown many advances, especially in mouse models of human diseases, and including the formation of various specialized cell types and even the generation of entire organs in vitro and possible genetic rejuvenation of aging muscles in mice in vivo," the study says. "Also, recent studies have put more emphasis on the influence of the stem cell niche, i.e., the physical microenvironment where stem cells reside in the organism, studied so far mainly in Drosophila and vertebrates, but noted also for sponges and flatworms. This will be an important consideration in future studies that utilize simple animals as models of stem cell function in human aging and disease. Thus, the knowledge gained from studying the simplest animals may someday provide keys for the extension of healthy human life to ages far beyond what is possible today."

Beyond human lifespan, people have also found inspiration for other advancements through nature. For instance, the development of the airplane. Throughout all of human history, men and women have looked to the skies and envied the flight of the birds. When it came time for people to recreate this mobility in the form of aircraft, it was the birds that provided much of the mechanics. According to Smithsonian Magazine, "Notes and inspiration from bird flight were integral in helping people finally achieve the goal" of human flight.[2]

---

and Longevity in the Simplest Animals and the Quest for Immortality," *National Library of Medicine* 16 (July 2014), 66–82, https://doi.org/10.1016/j.arr.2014.05.003.

2. Erin Malsbury, "How We Lifted Flight from Bird Evolution," *Smithsonian Magazine*, December 17, 2020, https://www.smithsonianmag.

"Through inspiration and careful observation of these marvels, humans have used bird flight to advance the goal of taking to the skies in our own unique way," the magazine reported before elaborating on the history of humans mimicking birds in an attempt to reach the skies. "Throughout history, humans dreamed of and attempted to fly by simulating the flapping wings of birds. In Greek mythology, wings made of wax and feathers carry Icarus through the sky and eventually too close to the sun. In 1505, Leonardo da Vinci wrote a codex on the flight of birds and designed—but never built—a flapping device for flight called an ornithopter."

It continues:

"Watching birds glide inspired inventors to build gliders with stationary wings that have a curved shape to create lift. Engineer Otto Lilienthal pioneered the first real attempts to fly by studying birds intensely, even publishing a book in 1889 called 'Birdflight as the Basis of Aviation' … People used soaring birds as early guides for creating lift and aerodynamic shapes, but before the Wright brothers, no one had figured out how to effectively control movement in larger aircraft. Lilienthal directed his gliders by shifting his body weight back and forth—similar to modern hang gliders—but the method was dependent on the size and weight of the aircraft … One of their keys to safe soaring—and steering—was a concept called wing-warping, which is mirrored in bird flight. By changing the wind-facing angle of the wings on one side, the Wright brothers realized you could increase the lift on one side, causing it to rise and turn the aircraft.

---

com/blogs/national-museum-of-natural-history/2020/12/17/how-we-lifted-flight-bird-evolution/.

They confirmed this idea by watching vultures glide in circles."

"Learning the secret of flight from a bird was a good deal like learning the secret of magic from a magician," Orville Wright said. "After you once know the trick and know what to look for, you see things that you did not notice when you did not know exactly what to look for."

Natural Lessons

Since it is often helpful to take these lessons from the natural world, as the Wright brothers demonstrated, let's look at how nature can teach us about various virtuous attributes.

Adaptability

Adaptability is one of the most important factors for survival for all animals, including humans. In the natural world, adaptability has been thought to be at the heart of Darwin's theory of evolution by natural selection. It's also the way so many wild animals have continued to survive and even thrive in a world that has dramatically shifted around them, including due to human intervention. In the world of business and innovation, this skill of adaptation is also incredibly crucial.

Writer and product expert Kyle Evans has addressed this intersection, noting, "In nature, there are a variety of adaptations that organisms use to respond to changes."

"Adaptability has always been among my top traits for career success, whether for product managers and developers, or for anyone working in technology and beyond," Evans wrote. "In nature, this is the ability for a plant or animal to change its own characteristics or features to better adapt to an environment. Like a bird developing a long beak. For us, we don't necessarily need to change our own physical characteristics to adapt. But we can create other structural or physical adaptations, for us, for our teams, or for our companies. We can change and adapt the environments we work in to the processes that we follow."[3]

Patience

There is a proverb dating back to the fifth century that says, "Patience is a virtue." As you probably know or have gathered, this means that patience is a positive quality to demonstrate in your life. Patience is also something that we can learn from the natural world. Lao Tzu, an ancient Chinese philosopher, said, "Nature does not hurry, yet everything is accomplished." Elaborating on that quote, Traditional Chinese Medicine World Foundation wrote, "In our move-faster-so-you-don't-lose-out society, many of us seem to have let our patience slip. We've become accustomed to getting immediate responses via email and text, and as a result, our patience for the slower side of life has gone by the wayside. We forego brick-and-mortar stores for shopping online and push many things on our to-do lists to the last possible minute. Humans are stressed. But Nature is truly our best teacher, if we take the time to relax and observe."[4]

---

3. Kyle Evans, "The Importance of Adaptability: A Changing World Requires a Changing Mindset," *Prodity*, August 23, 2023, https://www.productthinking.cc/p/the-importance-of-adaptability.
4. "Nature Knows: Patience," Traditional Chinese Medicine World Foundation, https://www.tcmworld.org/nature-knows-patience/.

The foundation continued, asking, "If we slow down and take time to appreciate Nature, what can we learn?"

"Watch the sky as day turns to night and then into day again. Observe the ocean as waves wash in and wash out. Spend time watching birds busily build warm, safe nests before laying their eggs. And observe an apple tree as its blossoms patiently wait for bees to pollinate them in order to develop fruit," the foundation wrote. "Humans cannot intercept and change these patterns. Nature doesn't hurry. It knows that its patterns bring forth life. If we were to apply even an ounce of Nature's patience to our daily lives, what might change? Would we move slower and savor life's precious moments?"

The foundation also issued a related recommendation:

"This week, spend some time walking among the trees and flowers. Observe Nature's interactions and infuse its teachings into your life."

Resilience

In my mind, one of the best symbols of resilience on Earth is the rebirth of forests. The tropical rainforests are often referred to as "the lungs of the planet," because they draw in carbon dioxide and breathe out oxygen. An incredibly important part of regulating oxygen and climate change, the world's forests are constantly threatened by both natural and man-made wildfires. Yet they have proven to be extremely resilient by regenerating in the wake of devastating burns. The mechanism for this regrowth is through a process called ecological succession. According to the University of Chicago, "Ecological succession is the process by which the mix of species and habitat in an area changes over time."

"Gradually, these communities replace one another until a 'climax community'—like a mature forest—is reached, or until a disturbance, like a fire, occurs," according to a press release from the school. "Ecological succession is a fundamental concept in ecology. The study of succession was pioneered at the University of Chicago by Henry Chandler Cowles, who was also one of the founders of ecology as a discipline, as he studied the plants of the Indiana Dunes."[5]

Secondary succession, according to the school, "happens when a climax community or intermediate community is impacted by a disturbance."

"This restarts the cycle of succession, but not back to the beginning—soil and nutrients are still present," it added. "For example, after a forest fire that kills all the mature trees on a particular landscape, grasses might grow, followed by shrubs and a variety of tree species, until eventually the community that existed before the fire is present again."

This is just one of the numerous ways in which nature builds back after setbacks. This is something that we can think about amid our daily dramatic turmoil, and use as a means to grow back our forests in our minds.

Interconnectedness and Maintaining Balance

One thing that nearly every human being on Earth can agree upon, regardless of our differences, is that *we are all connected.*

5. Max Witynski, "Ecological Succession, Explained," *UChicago News*, https://news.uchicago.edu/explainer/what-is-ecological-succession.

Whether you believe that we were created by a deity 6,000 years ago (we weren't), or you understand that our species is hundreds of thousands of years old, and evolved from a common ancestor we share with apes, you likely agree that every single human on Earth came from the same place. And not only that, but you also probably believe that every single other living being and non-living entity also has identical, shared origins. People of all faiths believe that some form of deity created the planet and everything on it, whereas people with no faith, for the most part, believe we all originated as one singularity before expansion created just about everything in the universe. And this universal idea of connection between everything else on Earth is an important thing to remember day to day. If you remember that we are all part of the same force, you might be more likely to extend a hand to help a stranger, or less likely to pollute or otherwise destroy our planet.

Caregiving

Above all else, one thing we can learn from animals and nature in general, is how to care for something. Famed painter Bob Ross once said, "We need to take care of these little creatures because they're sort of the barometer to tell you how the ecology is doing and how we're taking care of this old world. The animals disappear, guess who's next?" This is *why* it's important to care for the other beings with whom we share this world. And we can learn more about *how* to accomplish this by looking at nature itself. There are numerous species of animals that have shown this trait, including lionesses. I find this to be a particularly apt example because, despite being correctly known as strong creatures and fierce predators, they are also known to take in and adopt animals to whom they did not give birth.

"There have been multiple instances of lionesses adopting young, orphaned animals. In 2002, a lioness in northern Kenya made worldwide headlines after she was spotted cuddling up to baby antelopes—a species she would normally prey upon. Within a few months, the caring lioness had 'adopted' five young antelopes and was seen lying down next to them to ward off any would-be attackers," the International Fund for Animal Welfare writes. "A similar story spread in early 2020 when a two-month-old leopard cub was adopted by a lioness in India, even though these two species are normally at odds. The lioness, who had her own cubs the same age, spent weeks nursing, feeding, and caring for the leopard until he died."[6]

We can learn a lot from the lioness and her adoptive nature, but looming even larger than that when it comes to caregiving is the instinct and need for us to take care of the Earth herself. An idea even expressed in Holy Books (Genesis 2:15: "The Lord God took the man and put him in the Garden of Eden to work it and take care of it."), taking care of our environment in a sustainable manner, is also something we can find in nature. For example, you could look to the beavers, who, in their quest to protect themselves and create the perfect environment, can actually help mitigate climate change by storing water, reducing flooding, and improving water quality.

Retired biologist Gina Melton said to me in an interview, "Animals in general keep me amazed and humbled."

---

6. "These Amazing Animal Moms Adopted Orphans as Their Own," International Fund for Animal Welfare, March 8, 2024, https://www.ifaw.org/journal/amazing-animal-moms-adopted-orphans.

"We share a lot of our DNA with other primates. Chimps, gorillas, elephants and dolphins and whales have complex social structures, culture, and altruism," Melton explained. "Pets have been good for my two adult autistic children. They offer silent comfort and friendship, teach compassion and empathy."

Later in our interview, she said, "Since I studied mammalian biology, I'm partial to them, and biologists were originally naturalists. Our connection with other living things keeps us more human, I think."

When it comes to evolution, Melton expressed, "it's far from done with all living things."

She continued, "We're evolving constantly. It's little things, like humans losing their wisdom teeth, us growing taller and heavier than our ancestors."

Conclusion: Next Steps

"Nature's music is never over; her silences are pauses, not conclusions."

—Mary Webb

So far we have talked mostly about *which* actions one might take to improve their lives, or, more specifically, we have discussed *how* to look at evidence and decide from there which actions would be best. But laying out the blueprint is one thing; constructing a stable frame and then building out a house is something beyond that.

In other words, there's a big difference between listing the steps needed to live an evidence-based life, and actually putting that into place.

The first action to take in order to use science and critical thinking to make your life better or easier, is simply to use it when evaluating anything you hear or that you're told, to make sure it's from a credible source and the content stands up to scrutiny. Entire books have been written on this subject alone, including one by me (*The Curious Person's Guide to Fighting Fake News*). But it's also a pretty simple concept: anything you see or hear should run through a filtration system to determine its veracity.

Establishing a Baseline

During a polygraph test, the subject is asked a series of non-controversial questions that allow the administrator to see what the normal levels would look like. This is what allows a lie to be distinguished from the truth when reviewing the data. Unfortunately, so-called "lie detector" tests are highly unscientific, and can be fooled by anyone who has the proper coaching. But the concept is relevant here, since this is what you can do with your life, and your understanding of what is true.

To begin with establishing your gullibility baseline, we should start with the gullibility scale, developed by Macquarie University's Alessandra Teunisse and her team. Susan Krauss Whitbourne, Ph.D., a Professor Emerita of Psychological and Brain Sciences at the University of Massachusetts Amherst, has written about this exact subject. As she states, gullibility itself goes "beyond a lack of critical thinking."

"In gullibility, you succumb to a false premise, such as that belief a Nigerian prince is going to make you rich. Teunisse et al. believe that gullibility is an enduring personality trait that make people more likely to become victims of all kinds of scams. Even in the face of potential cues suggesting they're being scammed or fooled, gullible people show a type of 'deception blindness' to cues that someone else is untrustworthy. Furthermore, their gullibility can't be explained by situational factors, such as being tired, distracted, or overwhelmed with requests and offers that clog up your email inbox. Gullible people, the Australian authors propose, are low on the social intelligence that allows them to sniff out a fake offer, request, or even potential romantic partner's interest in you," Whitbourne wrote. "People who are gullible may additionally, according to Teunisse et al., be highly responsive to persuasion tactics, such as the pressure of a charming and convincing salesperson. How many times have you been led to buy something you didn't really need because it sounded like you were getting a fantastic offer? In your relationships, were you ever talked into revealing more about yourself than you intended to by an ingratiating potential partner? Later on, you regret having given in to these people because neither the deal nor the partner were as good as they sounded in the heat of the moment. It's not that the offer was necessarily based on false pretenses, but that you became pressured by their highly manipulative tactics into a decision that was not exactly ideal."[1]

This is a test you can perform yourself. You simply have to rate yourself 1–7, from strongly disagree to strongly agree, on the following 12 statements:

1. I'm pretty good at working out when someone is trying to fool me.
2. I'm usually quick to notice when someone is trying to cheat me.
3. I'm pretty poor at working out if someone is tricking me.
4. I quickly realize when someone is pulling my leg.

1. Susan Krauss Whitbourne PhD, ABPP, "A New Way to Test Just How Gullible You Really Are," *Psychology Today*, March 7, 2020, https://www.psychologytoday.com/us/blog/fulfillment-any-age/202003/new-way-test-just-how-gullible-you-really-are.

5. It usually takes me a while to "catch on" when someone is deceiving me.
6. I'm not that good at reading the signs that someone is trying to manipulate me.
7. My family thinks I am an easy target for scammers.
8. If anyone is likely to fall for a scam, it's me.
9. My friends think I'm easily fooled.
10. Overall, I'm pretty easily manipulated.
11. People think I'm a little naïve.
12. I guess I am more gullible than the average person.

Whitbourne further explains how to use your 1–7 rankings to get your gullibility quotient:

"To score yourself, reverse your ratings on numbers 1, 2, and 4 so that the 'strongly agrees' become 'strongly disagrees.' Now, calculate your total on items 1 to 6 to measure your Insensitivity to Untrustworthiness, and items 7–12 to score your Persuadability. On average, participants scored between 16–17 on Persuadibility (i.e. toward the low end of the scale) and slightly higher on the Insensitivity scale, between 18–19. If your scores were above 22 to 23 on these two scales, this suggests that you may want to turn up your antenna when others are trying to lure you into a trap."

To take that even further, a company called IDRlabs created what it calls the "Gullible Person Test," which is based on the Gullibility Scale developed by Teunisse and her research associates. The online test isn't a clinical assessment, but can still be helpful in establishing your belief baseline.[2]

2. Try the Gullible Person Test: https://www.idrlabs.com/gull-

Once you know how gullible you are, or more accurately, how easy it is for you to personally accept suspect information without credible validation, it's easier to know how to adjust your perception filter. And imagine what would happen if every single person paid special attention to their gullibility and to each extraordinary claim with which they are presented. Think of all the money that could be saved by avoiding scams and bad deals more generally. For example, the Federal Trade Commission (FTC) estimates that hundreds of thousands of people fall victim to pyramid schemes each year. In fact, in 2022, the FTC announced it was returning more than $149 million to AdvoCare distributors who lost money as a result of the health and wellness pyramid scheme. That's almost $150 million from *one scheme* that could have been prevented if more people were actively considering where they put their money.

Taking Action

Adjusting your truth filter and ensuring you're only taking in the best information is an amazing first step, but it's still just the beginning. Once you know you're dealing with the best data possible in all areas of your life, the next step is to take some action.

All the facts in the world can't help you, if you don't help yourself. A lot of self-help and motivational works emphasize thinking about your goals and keeping positive vibes, but that works to the detriment of the individual reader. Someone who is just trying to make their life better needs to know that the best

ible-person/test.php.

way to succeed is not to wish for something, but to *act on it.* Yet at the same time, it's important to note that a positive attitude *can* help you be happier as you move along that path.

Whether your goal is to improve your neighborhood, to cure cancer, or simply to get a better job, the answer won't be found by posting on social media. Results come not from thoughts and strong emotions, but by stepping up and doing the hard work. Whether that takes the form of physically cleaning up your streets or merely supporting elected officials who value the same end goals as you do, the key is to avoid inaction.

The Power of Writing

One thing to consider as you work to make your life better through good decisions is how you start your day. For instance, Amanda Reill, an author, speaker, coach, and entrepreneur, has given one example of something that helped her be more productive at work: journaling.

Specifically, Reill said that she read Julia Cameron's book, *The Artist's Way*, and heard about "morning pages"—defined as "three pages of journaling on paper, completed first thing after you rise."

"The rules surrounding morning pages are loose. You can write anything that crosses your mind: a dream you just had, your fears about the day, gratitude, or even 'I don't know what to write' over and over again. While the practice was originally designed to help artists break out of a creative rut, it's a technique that proves valuable for every aspect of life. After committing to it for several months, I realized it was positively influencing how I organized my thoughts and helping me make better, more thoughtful decisions at work," wrote Reill, who told a story about being called into a meeting by Human Resources while worrying about losing her job.[3]

"I needed courage. When I hit the pages that morning, I confronted the fears I had about how this meeting would go. I had some decisions to make about how I would respond if things went south," she added. "During the meeting, I spoke clearly, calmly, and with confidence. I did not ramble on out of nervousness. I had already decided who I was going to be that day, regardless of how the conversation went. After leaving the meeting, I was actually shocked at how easy it had been to advocate for myself. I quickly connected this pleasant surprise with the day's writing practice. Because I had taken the time to check in with myself about what I thought, and commit to it by writing it on paper, it transformed how I felt about the conversation and how I handled it in the moment."

This anecdote is backed up by substantial research. Not only is it helpful to get your feelings out on paper, a practice that can help you prepare for the day's challenges, but the physical act of handwriting itself has actually been shown to help to train your brain.

"Adults studying new symbols, such as Chinese characters, might enhance recognition by writing the characters by hand, researchers say. Some physicians say handwriting could be a good cognitive exercise for baby boomers working to keep their minds sharp as they age," according to a report from the Wall

3. Amanda Reill, "A Simple Way to Make Better Decisions—The Case for a Daily Writing Practice," Reill Coaching, https://www.reill-coaching.com/post/daily-writing-practice.

Street Journal. "Recent research illustrates how writing by hand engages the brain in learning. During one study at Indiana University published this year, researchers invited children to man a 'spaceship,' actually an MRI machine using a specialized scan called 'functional' MRI that spots neural activity in the brain. The kids were shown letters before and after receiving different letter-learning instruction. In children who had practiced printing by hand, the neural activity was far more enhanced and 'adult-like' than in those who had simply looked at letters."

Examples from Daily Life

When we are talking about putting theories into practice in an effort to make your life better, the best place to start is with those who have already lived it. In doing the research for this book, I spoke with thousands of individuals about how they use science and critical thinking to make their lives better or easier.

Some said they use both science and reason extensively when evaluating what they hear, to make sure it comes from a credible source and stands up to scrutiny. Others, on the other hand, put these practices into place when it comes to their work.

One educator said, "As a teacher, I use it every day as I educate my students. I hope that I've passed on the skill to them so that future generations will be better prepared to navigate the world than we are currently."

Another person who focused their reply on their occupation also chimed in, saying, "I'm a woodworker," before adding, "The shapes I cut for joinery, the adhesives I use, and of course the finishing process all rely heavily on scientific principles."

Similarly, a welder/fitter/foreman at a welding shop said, "My daily activities include Pythagorean theory (to make sure that things are square and not just a parallelogram). My co-worker insists that I have to know trig, but I don't use it often enough. Math is pretty much all I do all day. When I tell people what I do, it's usually 'math.' It usually comes [as] a surprise too because I work in a trade."

A professional massage therapist said, "Every single day I use anatomy, physiology, pathology, nutrition, and even pharmacology."

Some also acknowledged that there is a downside to being properly informed. Invoking the popular proverb, "Ignorance is bliss," one individual told me, "If I could just believe any old thing, I'd probably be happier." But that can only take you so far. Perhaps in some ways you might be happier being less educated and less informed, but you'd also be less successful and less able to properly address glaring problems in your own life.

One person who has a mast cell disorder explained how she uses science and reason to separate woo-based "treatments" from legitimate medical therapies that could help her. This could potentially be a life-saving result from an evidence-based approach to health.

Along those same lines, many people told me they use the scientific method to properly treat their ailments. One individual said, "I really started consciously using the scientific method once diagnosed with PTSD. I used it as a form of CBT when I could feel the panic starting to rise."

Similarly, a bioengineering student told me they use science "every day when I rationalize certain reactive emotions."

"It helps me to better understand my interaction with the

world and people around me," they told me.

Some people actually stated that they use critical thinking to tackle problems that haven't even arisen yet. One person said, "The most prevalent way I use critical thinking is in my extrapolative faculty."

"Finding patterns, and extrapolating their possible consequences allows me to foresee, and find solutions for problems before they even occur," they added.

Others talked to me about how an evidence-based approach to life helps them to avoid the "draining" rituals associated with superstition or supernatural beliefs.

"Life is easier in general when your mind is free to seek out scientific explanation for any number of 'mystical' subjects," one person told me. "This definitely makes life easier, because irreligious people do not need to perform rituals or bear the burden of superstitional behavior."

Some people were more focused on their children than themselves.

"I have a 7-year-old full of questions about EVERYTHING. In any given week, we'll cover anything from the condensation cycle to what causes stars to be different colors to why roller coasters don't fall off the track when they go through a loop," they said.

Making Life Safer

Improving your life and achieving success are worthy goals, but what do they *really* mean to you if you aren't staying safe to enact that strategic blueprint? One of the biggest ways science and reason can impact our lives is through evidence-based crime mitigation strategies, which can literally save your life.

On a macro level, evidence-based crime prevention has been addressed by the United Nations Office on Drugs and Crime, which stated that, "Preventing crime requires concerted and well-coordinat[ed] efforts among the government and other relevant stakeholders and there are many theories and practices for crime prevention which can work in one situation but fail in another."[4]

"Evidence-based policymaking is essential, in particular in the context of the complexity of crime prevention as well as the overarching efforts and Goals in the 2030 Agenda," according to its latest report at the time of this writing. "Evidence-based crime prevention aims at identifying the factors which are associated with different types of crime leading to the selection of a set of strategies and programmes which can influence the underlying factors to prevent crimes and help policymakers with correctly targeted actions."

But what I'm talking about is evidence-based crime prevention on the micro, or personal, level, and the definition is simpler: using strategies that are proven by science to keep you and your family and your community safe.

One example of this is through a home alarm system. If you're looking to improve the safety of your home, it helps to look at the evidence, and the best place for that is often a solid,

4. "Evidence-Based Crime Prevention: Statistics, Indicators and Evaluation in Support of Successful Practices," United Nations Office on Drugs and Crime, March 2021, https://www.unodc.org/documents/commissions/Congress/workshops/concept_notes/Concept_Note_Workshop_1.pdf.

peer-reviewed scientific study. In this case, one of the best studies available is from Rutgers University School of Criminal Justice. The school says its researchers used five years of data to conclude that "residential burglar alarm systems decrease crime."

"While other studies have concluded that most burglars avoid alarms systems, this is the first study to focus on alarm systems while scientifically ruling out other factors that could have impacted the crime rate," the school announced at the time of the study's publication.

The study not only found that individuals and families are kept safer by home alarm systems, but also actually concluded that they impacted the total crime levels of the entire city.

"The study credits the alarms with the decrease in burglaries and the city's overall crime rate," said the study's author, Dr. Seungmug (a.k.a. Zech) Lee. "By using sophisticated in-depth research techniques, we were able to eliminate the variables that impact crime rates and focus directly on the impact alarm systems have on residential burglaries."[5]

Just like when you go to the doctor, it's usually a good idea to look for a "second opinion" when it comes to your search for evidence-based actions. However, in this case, we aren't looking for any opinions at all. Still, it's important to verify the other findings.

In this case, we can take a look at the National Criminal Justice Reference Service to see a study entitled "Alarm system effectiveness research." That study, sponsored by the Multnomah County Department of Public Safety in Oregon, concludes that

---

5. "Rutgers Study Finds Alarm Systems Are Valuable Crime Fighting Tool," *Rutgers*, February 5, 2009, https://www.rutgers.edu/news/rutgers-study-finds-alarm-systems-are-valuable-crime-fighting-tool.

alarm systems "are a valuable method of crime prevention."[6]

Not only do alarm systems help individuals, neighborhoods, and cities, but even the police department itself benefits from the implementation of alarm systems, according to that study, which concluded that "police spent considerably more investigative time with non-user victims."

This is also a situation where science can save us money because, according to that study out of Oregon, "certain types of alarm systems lowered dollar loss" amid break-ins.

Try Remaining Positive

There is a lot of hate and anger in the world today, but that's likely always been the case. It may not be a unique issue, but it's definitely one we can work to improve together. If someone upsets or frustrates you, remember that you are both made of the same materials and you're both simply looking for answers in this crazy life. After all, on a cosmic level, we are all just tiny organisms on a rock barreling through space. So when we find ourselves at each other's throats over things that aren't all that important, it's crucial to remember: We can do better. We can give each other the benefit of the doubt in some cases.

When interacting with lots of people in daily life, it can be difficult to remain positive. This is especially tough when engaging in controversial debates about politics, religion, or even philosophy more generally. But we should keep in mind that this

6. M. A. Y. Rifai, "Alarm System Effectiveness Research," *NCJRS Virtual Library* (1977), https://www.ojp.gov/ncjrs/virtual-library/abstracts/alarm-system-effectiveness-research.

hypothetical person with a different perspective, though they have opposing views, is still a person. This "opponent" of yours may have faced issues that you didn't face in life, and they may have something to teach you—even if neither of you recognizes it. Anything can be a learning experience if you make it one.

But *why* remain positive? After all, positivity is commonly encouraged by self-help works that are based more on spirituality than on science. Is there really proof that this is helpful?

One of the best parts about following the evidence wherever it leads is that we can get good ideas from anywhere, as long as they are indeed valid. And in this case, positivity can certainly be one of the keys that drives success. But it's not enough just to believe that being positive is going to change our lives. In reality, positivity is only one aspect of the broader picture, and it's important to look at the science to determine when it is helpful.

As it turns out, positivity has the strongest impact when it comes to productivity in group settings. If you are working with numerous other people, whether it's in an office or on a project in your community or anything else, a positive mindset is a tool in your toolbox as you seek to get the most out of the experience. According to the University of Pennsylvania, "positive connections with other people were literally wired into our brains for survival and are integral to happiness."

"A recent study by the University of Kent found that strong bonds with family, friends, and co-workers can help boost physical and mental health and overall well-being. As you've likely experienced firsthand, both positivity and negativity among colleagues in the workplace can be contagious," according to a write-up published by the school. "In work environments where employees lack strong relationships, morale will inherently suffer, but when colleagues share strong bonds, it can reduce stress, improve collaboration, and boost productivity. That's why participating in team-building exercises and events where colleagues can connect and build rapport is so vital."[7]

Taking This into the World

You make an average of 33,000 to 35,000 total decisions each day, according to the Harvard Business Review. While it's impossible to take a thoughtful, reason-based approach to every single one of them, it *is* worth considering exactly how many of those choices are made based on whims and without taking proper consideration. If, through thoughtful practice and careful deliberation, you are able to change how you make just 10% of your daily decisions, that is thousands of actions you'll be taking every day that bring you closer to living an evidence-based life free from irrational decision-making.

Since the bulk of our decisions are micro-choices often made without thinking, the first step should be simply thinking about them more. But what about the bigger questions in life, such as how to achieve a major goal? For that, it comes down to something even simpler: taking action. Whether it's paralysis from an overabundance of items on your to-do list or you've simply been caught up in a purported spiritual life hack, there is a tendency for many people to *not act*. Some people can't make

7. "The Science of Happiness at Work: How Positive Psychology Can Increase Productivity," *Penn LPS Online*, October 6, 2023, https://lpsonline.sas.upenn.edu/features/science-happiness-work-how-positive-psychology-can-increase-productivity.

the needed moves because they are too overwhelmed, while others believe they can get what they want through prayer or some other petition to the supernatural. Either way, the hope is that this work can help you on your steps forward. After all, if you need a car, do you go through the hard work of researching and financing and test driving, or do you merely put a picture of a vehicle on a vision board? Personally, I know what I would do. It is also important to note that taking action is incredibly important, but not all actions are good. The best way to make sure that you are taking positive steps is to utilize the scientific method, where applicable.

As you go through life, you'll be tempted by purported shortcuts to what you want. For example, some might say all you have to do is pray, and God will make your life better. But there is no substitute for doing hard work. Faith is something that *can* help some people on a personal level. Studies have shown that humans have an innate desire to explain the unknown, and it can also be helpful to believe you are going to paradise after this life, so taking these major questions off one's plate can be helpful to some. Yet it's also important to acknowledge that, day to day, religion might not help you in achieving results and meeting goals. For that, it takes dedication to take the steps needed in the natural world.

Improving your life typically works from the bottom up. If you start looking thoughtfully at your thoughts and feelings first, that can have a positive impact on your mindset and mental health. In turn, that positive change in how you *think* can affect the small decisions you often make without any consideration. Those can then influence the bigger decisions that are directly responsible for success or failure in achieving your goals. This is one way that evidence-based decision-making can help you get what you want out of life. Another way to succeed is to boost your financial well-being. Once again, this happens from the bottom to the top. It's not *always* about changing careers or even being promoted, although those both can come along with a consistent dedication to thoughtfully considering your position in life. Often, financial improvements start at home with micro-decisions. Even something as small as how you load your dishwasher can affect your utility bill, which in turn can change your life when it comes to monetary security. This is why it's important to consider as many of your day-to-day decisions as possible in such a way that you can change your relationship to money.

Some readers might think that data-driven decisions and reason-based living take the fun out of this life, but I think that balance is key. You don't want to make every single decision based solely on a crude analysis, or, of course, you will live life like a robot, but at the same time, you can improve the quality of life you share with your loved ones if you're able to think carefully about your minor daily decisions. The truth is that thinking more carefully about things, and even understanding them on a deeper level than you may have in the past, can actually bring more awe into your life, if done properly. For example, you could free up your mind from little stresses with which it was once consumed, which could then help you open yourself up enough such that more positive things can enter your life—maybe even love.

As important as it is to achieve your dreams in the worlds of your career and finance, it's equally important not to forget that the base underpinning of that is improving your own mental health. Managing your emotions and personal relationships is another key to enjoying life, and it should not be ignored. The good thing is that many of the same principles that go into maximizing your financial outlook can also be applied to anything else, including managing emotions. Then you get to reap those benefits twice. You work hard on your emotional well-being, and then you start to make better decisions in other areas of life, as well, as a result of that.

Whether you are working to better yourself internally or to achieve career success, one of the best methods to make it happen includes utilizing science. For the typical person, however, science can seem out of reach, or perhaps even irrelevant. The layperson hears "science," and they think about beakers and lab equipment. They think scientific evidence isn't something they care about when it comes to living their daily lives. In fact, many of them might have even convinced themselves that science is out of reach for anyone who isn't a legitimate scientist. I actually met someone like this, we will call her Arizona, who didn't understand her potential when it came to scientific inquiry. I told Arizona that I was a science writer and data advocate, and her eyes just glazed over. She instantly said, "Science is boring, sorry." I didn't hold this against Arizona. She was an interesting and curious person, so I just figured science didn't interest her, and that was OK. But when I asked her what *did* interest her, her reply ended up shocking *me*. She told me she loved to play with, make, and test various skin care products. When I asked her to tell me more about it, she said she would try skin care products that have anecdotal backing, but which likely haven't been tested on a mainstream level. She said she would test each of them, see how her skin reacts, and then review them and share videos with her followers. When I broke down the scientific method for her, step by step, she was stunned. It turns out, just about *anyone* in the world is capable of carrying out the process by which we determine what's true.

I think we should all remember Arizona's story next time we think of science as something only for school, or something only available to a class of people called scientists. We should realize that science is the best method humans have for deciphering falsehoods from reality, so why not use it when trying to better ourselves? At the same time, it's important to keep in mind that science isn't the *only* lure in our tackle box. Logic and reason are similarly key to making sure the steps we are taking are actually effective, as is critical thinking more broadly. In fact, a journey toward self-improvement requires all these and more. If we are fishing for new ways to improve our lives, we might not only need more than one lure; we might need several other types of bait, as well.

The fact is that the source of our inspiration isn't what matters. Whether you are motivated to better yourself because you identify with the independence of a lion, because you've created a pros and cons list, or because statistics say you should do so if you want to accumulate wealth, the important thing is that you take the steps needed. The longer you put it on the back burner or wait for karma or fate or God to step in, the more time passes without you being your best self. And that's something no one wants.

About the Author

David G. McAfee is author of *Hi, I'm an Atheist, No Sacred Cows, Disproving Christianity, Atheist Answers, The Belief Book,* and numerous other published works. He lives in the High Desert of Southern California.

www.ingramcontent.com/pod-product-compliance
Lightning Source LLC
LaVergne TN
LVHW101915280826
846781LV00020B/85